SpringerBriefs in Genetics

For further volumes:
http://www.springer.com/series/8923

Pete Humphries · Marian M. Humphries
Lawrence C. S. Tam · G. Jane Farrar
Paul F. Kenna · Matthew Campbell
Anna-Sophia Kiang

Hereditary Retinopathies

Progress in Development of Genetic and Molecular Therapies

 Springer

Pete Humphries
Ocular Genetics Unit
Institute of Genetics
Trinity College Dublin
Lincoln Place Gate
Dublin 2
Ireland

Paul F. Kenna
Ocular Genetics Unit
Institute of Genetics
Trinity College Dublin
Lincoln Place Gate
Dublin 2
Ireland

Marian M. Humphries
Ocular Genetics Unit
Institute of Genetics
Trinity College Dublin
Lincoln Place Gate
Dublin 2
Ireland

Matthew Campbell
Ocular Genetics Unit
Institute of Genetics
Trinity College Dublin
Lincoln Place Gate
Dublin 2
Ireland

Lawrence C. S. Tam
Ocular Genetics Unit
Institute of Genetics
Trinity College Dublin
Lincoln Place Gate
Dublin 2
Ireland

Anna-Sophia Kiang
Ocular Genetics Unit
Institute of Genetics
Trinity College Dublin
Lincoln Place Gate
Dublin 2
Ireland

G. Jane Farrar
Ocular Genetics Unit
Institute of Genetics
Trinity College Dublin
Lincoln Place Gate
Dublin 2
Ireland

ISSN 2191-5563
ISSN 2191-5571 (electronic)
ISBN 978-1-4614-4498-5
ISBN 978-1-4614-4499-2 (eBook)
DOI 10.1007/978-1-4614-4499-2
Springer New York Heidelberg Dordrecht London

Library of Congress Control Number: 2012940944

Printed on acid-free paper

Springer is part of Springer Science+Business Media (www.springer.com)

Preface

The euphoria we experienced in 1989, in localizing 'the' gene for autosomal dominant retinitis pigmentosa to the long arm of chromosome 3, was an exuberance that was exceedingly short-lived. It immediately became apparent, through lack of evidence for genetic linkage in a second family, that 'one more' gene remained to be identified. The reality today, is that for this one disease, we are dealing with an immensely complex set of molecular pathologies—considering all hereditary forms of retinopathy, of which RP is the most common, loci for well over two hundred genes have been revealed, and the chances are that the same number again still remain to be identified. It is a demanding enough endeavor to develop a gene-based medicine for a disease in which only a single gene has been implicated, but in considering hereditary degenerative retinopathies, notwithstanding the triumphant successes such as those achieved for restoration of vision in those forms of Leber congenital amaurosis with mutations within the RPE65 gene, the question is: just how logistically and economically feasible is this? If our armament resided only in gene therapy, we have embarked upon a protracted journey. However, in spite of the immense genetic complexity facing us, commonalities are evident in the molecular mechanisms through which vision is lost in these diseases, and much evidence is now available to encourage us to believe that a combination of gene-based and molecular medicines will, inevitably, result in the development of effective therapies for many of these conditions within a realistic timeframe.

Contents

Contents

Chapter 1
Introduction

Of the almost 200 million current world-wide cases of visual handicap, syndromic and non-syndromic hereditary degenerative retinopathies involving progressive degeneration of photoreceptors, followed usually by more extensive degeneration of the retina, figure prominently among young and working aged people in the developed world [1]. While many discrete clinical entities have been described, including for example, rod-cone and cone-rod dystrophies, congenital nyctalopia, macular dystrophies, retinoschisis, optic atrophies, hereditary optic neuropathies, Bardet-Biedl and Usher syndromes etc., retinitis pigmentosa (RP) is the most prevalent cause of registered visual handicap among those of working age in developed countries [2–6], while Leber congenital amaurosis (LCA) is the most prevalent congenital cause of registered visual handicap [7]. Typical fundus features of RP (thinning of the retina, pigmentary disturbance and attenuation of retinal vessels) are illustrated in Fig. 1.1. RP is a progressive disease, of variable age of onset, rarely showing congenital signs of retinal degeneration, although in some cases, nyctalopia (night blindness) is evident in very young infants. Retinal function is routinely monitored using the electroretinogram, where corneal electrodes measure light-induced electrical impulses generated by the photoreceptors (the a-wave) and by second order retinal neurons (the b-wave) (see Figs. 1.2 and 1.3). LCA, as the name implies, is a congenital disease, and often there are no signs of retinal pathology in early fundus examinations, although photoreceptors, while maintaining viability, are non-functional (see fundus photograph, Fig. 1.4). LCA segregates in families as an autosomal recessive condition, whereas RP is either autosomal recessive, dominant or X-linked recessive in its mode of inheritance, although more complex forms of segregation, e.g. digenic, or tri-allelic hereditary patterns have occasionally been observed [8, 9]. As the human genetic linkage map developed, progressively populated by markers based on variability in the size of restriction fragments of DNA—so called restriction fragment length polymorphisms or RFLPs (later to be superseded by much more informative makers based on variability in the length of short microsatellite sequences, which could be analyzed using the polymerase chain reaction—a great advantage compared with the very cumbersome, though elegant technique of Southern blotting for analysis of

P. Humphries et al., *Hereditary Retinopathies*, SpringerBriefs in Genetics,
DOI: 10.1007/978-1-4614-4499-2_1, © The Author(s) 2012

RFLPs) it became clear that a sufficiently large number of genetic markers was becoming available such as to enable realistic systematic screening of the genome in the search for disease loci. One of the first early successes in genetic linkage mapping using modern genetic markers came with the localization of the first retinopathy gene by Bhattacharya and colleagues in 1984 [10], who demonstrated linkage between the RFLP, L1.28 and a locus for RP on the short arm of the X-chromosome. This seminal work stimulated similar research all over the world and in 1989, a study was reported in which linkage had been established between the genetic marker C17 (D3S47) on the long arm of chromosome 3 and the disease locus in a large family from Ireland in which up to 50 individuals were afflicted by a particularly early onset, though slowly progressive form of RP that segregated in the family in an autosomal dominant mode [11]. Anecdotally, parents of very young children were often able to diagnose the condition by observing the way in which the child behaved in dimly lit conditions owing to the presence of a congenital night vision defect. The large size of this family resulted in the generation of exceedingly powerful linkage data, a lod score of 14.7 with zero percent recombination between the marker and the disease locus—the largest lod score ever attained in a genetic linkage analysis of a single pedigree and in a subsequent multilocus analysis the lod score increased to approximately 20 [12]. The data pointed to a single mapped gene in the interval—the gene encoding rhodopsin. Could a dominant mutation within the rhodopsin gene, encoding the primary component of the visual transduction cycle, in the heterozygous state, be responsible for disease pathology in this family, and if so how? Shortly thereafter, the first rhodopsin mutation was reported by Dryja et al. [13] in a patient with adRP, a point substitution, Pro23His, located toward the intra-discal N-terminus of the protein. Interestingly, while this mutation turned out to be common in the USA, it was found to be absent in a study of up to 100 families with adRP in Europe [14], the mutation in the original pedigree from Ireland being an arginine to proline substitution at codon 347, a substitution within the 5th trans-membrane domain of the protein [15]. Today, over 120 mutations have been identified within the rhodopsin gene, largely in cases of adRP, although a small proportion of mutations cause congenital night blindness.

A second locus for a dominant form of RP was reported on the short arm of chromosome 6 in 1991 in a family with a later onset form of disease than that manifesting in the first pedigree and a mutation within the RDS-peripherin gene was found—a trinucleotide deletion, removing one of a pair of cysteine residues at position 118/119 of the protein [16]. Many mutations within the RDS-peripherin gene have to date been encountered in degenerative retinopathies, predominantly in autosomal dominant RP. RDS-peripherin is so-called because null mutations within this gene underlie a naturally occurring slow retinal degeneration in mice (rds-/-). Peripherin is a structural component located at the periphery of the outer segment disks of rod photoreceptors, and is also found in cone photoreceptor outer segments. Dominant mutations within the RDS-peripherin protein structurally destabilize photoreceptors leading to retinal degeneration.

Mutations within the rhodopsin and RDS-peripherin genes combined, account for about 30 % of cases of autosomal dominant RP. A major implication from

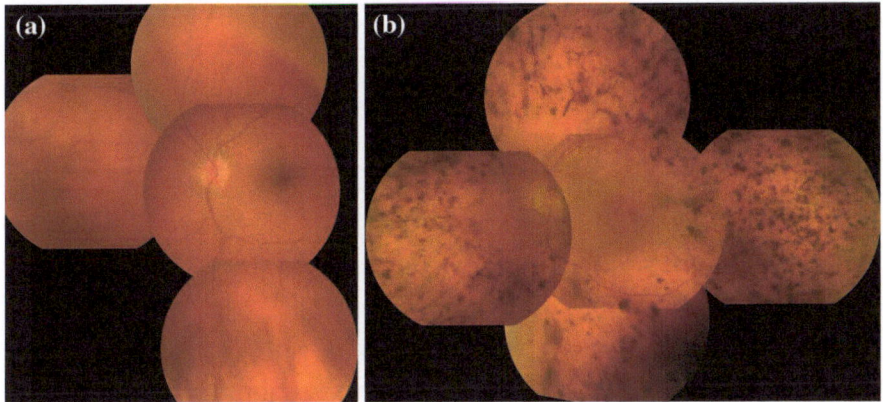

Fig. 1.1 Autosomal dominant retinitis pigmentosa caused by a mutation within the rhodopsin gene. Note the pale optic disk (**a**), attenuation of retinal vasculature and the widespread pigmentary deposits in the peripheral retina (**b**). There is also evidence of macular oedema

Fig. 1.2 Dark-adapted ERGs. *1* Nonrecordable rod-isolated response of an RP patient. *2* Rod-isolated response of a nonaffected individual. *3* Nonrecordable mixed road and cone response of an RP patient. *4* Mixed rod and cone response of a nonaffected individual. The time scale is 50 ms per division. The voltage scale is 500 μV per division

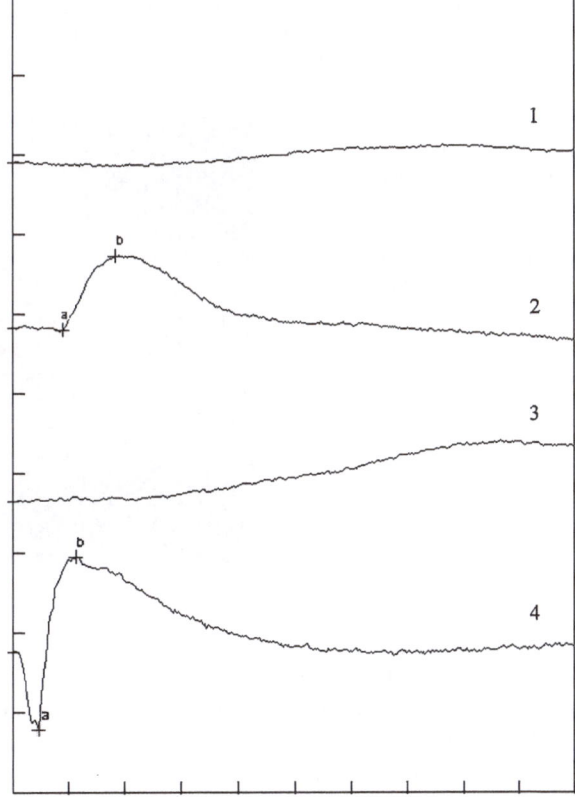

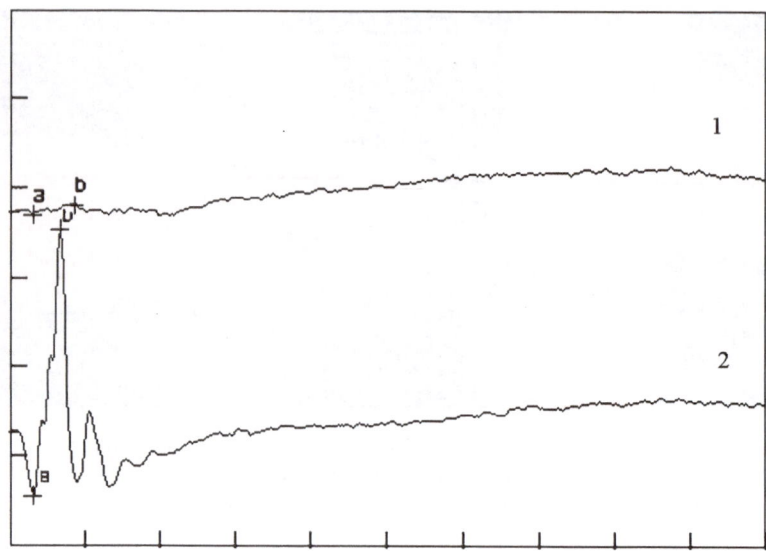

Fig. 1.3 Light-adapted ERGs. *1* Vestigial cone-isolated response of an RP patient. *2* Cone-isolated response of an unaffected individual. The time scale is 50 ms per division. The voltage scale is 500 μV per division

Fig. 1.4 Leber congenital
amaurosis: multiple fine
yellow/white dots are visible
in the periphery with atrophic
changes associated with
pigmentary disturbance
visible in the central macular
area

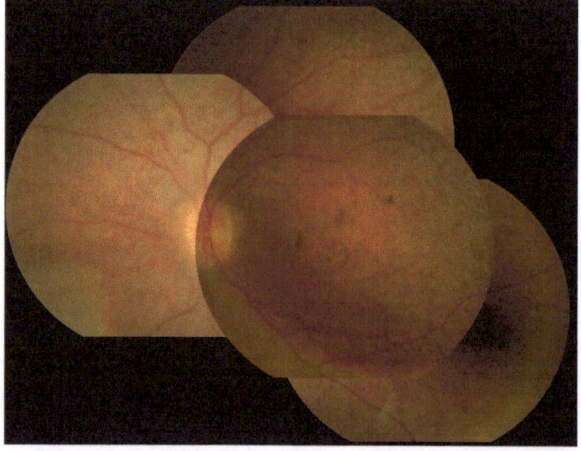

these findings, however, was that any other gene encoding a functional element of the visual transduction or visual cycle, or any other structural component of photoreceptors, could be considered as a disease-causing candidate.

A third locus for autosomal dominant RP was reported in 1993, following a genome-wide genetic linkage study on DNA from a Spanish family [17]. A mutation within the inosine monophosphate dehydrogenase gene (IMPDH1), encoding one of two rate-limiting enzymes of de novo guanine nucleotide biosynthesis was then identified, IMPDH1 being expressed in many tissues of the body and yet the mutated protein inducing disease symptoms only in retinal tissues [18].

Gradually, a picture of intragenic heterogeneity began to develop and in the ensuing years, RP and LCA turned out to be conditions of immense genetic diversity, we ourselves having described RP as possibly the most genetically heterogeneous of any hereditary condition in which molecular pathologies have so far been explored [19]. For autosomal recessive and X-linked recessive RP, autosomal dominant RP and Leber congenital amaurosis, mutations within 26, 19, and 15 genes have so far been characterized, respectively (these are listed in Tables 1.1, 1.2 and 1.3; for further information the reader is referred to [20]).

In some cases, the basis for photoreceptor degeneration caused by null or dominant-acting mutations within these genes has been well established. For example, the kinase MERTK, which is absent in a proportion of cases of auto-somal recessive RP, plays a role in assisting the RPE in phagocytosing shed rod photoreceptor outer segments disks [21]. Reduced ability to remove such debris from the retina (10 % of the outer segment disk membranes are shed into the RPE on a daily basis) suggests an obvious correlation between this genetic aberration and disease progression. In regard to autosomal dominant RP, a proportion of mutated IMPDH1 enzymes and an appreciable number of the 120 or so mutated rhodopsin proteins so far identified, misfold or aggregate and this can trigger apoptotic death [22–24]. RDS-peripherin is a critically important structural com-ponent of the outer segment disk membranes of rods and of the outer membranes of cone photoreceptors. Structurally, altered protein destabilizes disk- or outer-membrane structure, in turn compromising photoreceptor viability. However, in many cases the basis for disease pathologies induced by mutant proteins is more obscure, arguably perhaps the most interesting being those scenarios in which proteins are widely expressed, with disease pathology being limited only to retinal tissues. Included in the latter category, for example, are genes encoding at least three pre-mRNA splicing factors [25–31].

Upon this backdrop of complex molecular pathologies, an important question materializes: with such genetic heterogeneity, how realistic is it to envisage that preventive gene or molecular therapies can be developed for all genetic subtypes of RP and LCA and for the myriad of other much less common degenerative retinopathies? The answer, at least in terms of gene replacement for recessive disease, is, in principle, yes. In terms of gene therapy for autosomal dominant disease, the answer is again, probably yes, although suppression of dominant acting mutations at the transcript level, or at the level of the DNA itself, is a more demanding concept. However, developing and validating the large number of gene-based medicines required, together with mandated safety and toxicology studies for such a genetically diverse group of conditions, notwithstanding the fact that RP and LCA fall well within the category of orphan diseases, will be an exceedingly costly and time-consuming exercise. Parallel approaches, in which medicinal products are aimed at targeting molecular pathologies common to multiple genetic forms of disease, are therefore, clearly of great potential eco-nomic attraction.

Table 1.1 Genes identified in autosomal recessive and X-linked recessive retinitis pigmentosa (RP) and their function

Gene	Function
RPE65	Enzymatic component of the visual cycle in which all-trans retinal is recycled within the pigmented epithelium into the 11-cis form
ABCA4	A transmembrane protein transporter of outer segment disks
CRB1	Crumbs homolog 1; possibly a component of the inner segment scaffold of photoreceptors
USH2A	Usherin; a basement membrane protein of retina and ear
MERTK	A thymidine kinase, mutations within which induce defective phagocytosis of outer segment disk membranes
CERKL	Ceramide kinase-like protein, which may be involved in neuronal survival/apoptosis
SAG	Arrestin; a component of visual transduction cycle
RHO	Rhodopsin, the primary photoreactive pigment of rod photoreceptor cells
PDE6A	The alpha subunit of cGMP phosphodiesterase, a central enzymatic component of the visual transduction cycle
PDE6B	The ß subunit of rod cyclic GMP phosphodiesterase, a component of the visual transduction cycle
LRAT	Lecithin retinol acyltransferase, an enzymatic component of the visual cycle
TULP1	Tubby-like protein 1—with as yet unknown function
EYS	An extracellular matrix protein—one of the largest human genes at >2mb
TTPA	A protein involved in the transport of vitamin E
IRBP	Intra-retinol-binding protein, involved in the visual cycle
CDHR1	Protocadherin 21; a cell–cell adhesion protein
RGR	A component of the visual cycle—an RPE-specific protein involved in the conversion of all-trans retinal to 11-cis retinal
SPATA7	Spermatogenesis associated protein 7 which may be associated with transport in photoreceptor cilia
NR2E3	Nuclear receptor subfamily 2 group E3—a retinal transcription factor
RLBP1	Retinaldehyde binding protein 1, a component of the visual cycle
CNGB1	A component of the visual transduction cycle—ß subunit of rod cGMP-gated channel
PRCD	A protein of unknown function
IDH3B	NAD-specific isocitrate dehydrogenase 3ß—a component of Kreb's cycle
RPGR	A GTPase regulator
RP2	Homologous to cofactor C—mutations within this gene may cause defects in ß tubulin folding
C2ORF71	A protein of the cilia of photoreceptors connecting the inner and outer segments of the cells

Table 1.2 Genes identified in autosomal dominant retinitis pigmentosa (RP) and their function

Gene	Function
PRPF3	A pre-mRNA splicing factor 3
SEMA4A	A transmembrane semaphorin
SNRNP200	A small nuclear ribonucleoprotein
RHO	The primary component of the visual transduction cycle
RDS-peripherin	A structural component of photoreceptor outer segment disk membranes
GUCA1B	Guanylate cyclase activating protein 1B, a component of the visual transduction cycle
KLHL7	Kelch-like protein 7 of unknown function
RP9	PIM1-kinase-associated protein, pre-mRNA splicing
IMPDH1	One of two rate-limiting enzymes of de novo GTP biosynthesis
TOPORS	Topo 1-binding protein—its function as a retinopathy causing protein has yet to be fully elucidated
ROM1	A structural component of photoreceptor outer segment disk membranes which interacts with peripherin
NRL	A retinal transcription factor interacting with CRX
RDH12	Retinol dehydrogenase 12, an enzymatic component of the visual cycle
PRPF8	A pre-mRNA splicing factor
CA4	Carbonic anhydrase 4
FSCN2	An actin-bundling protein
CRX	A photoreceptor transcription factor
PRPF31	A pre-mRNA splicing factor
RPE65	Enzymatic component of the visual cycle in which all-trans retinal is recycled within the pigmented epithelium into the 11-cis form

Table 1.3 Genes identified in Leber congenital amaurosis (LCA) and their function

Gene	Function
RPE65	Enzymatic component of the visual cycle in which all-trans retinal is recycled within the pigmented epithelium into the 11cis form
CRBI	Crumbs homolog 1; possibly a component of the inner segment scaffold of photoreceptors
RD3	The RD3 protein associates with PML bodies in the nucleus
LRAT	Lecithin retinol acyltransferase, an enzymatic component of the visual cycle
TULP1	Tubby-like protein 1—with as yet unknown function
LCA5	Lebercilin; a protein of the connecting cilia of photoreceptors, interacting with other ciliary proteins
IMPDH1	One of two rate-limiting enzymes of de novo GTP biosynthesis
CEP290	A ciliary protein
RPGRIP1	RP GTPase regulator-interacting protein 1, which interacts with RPGR and localizes to connecting cilia of photoreceptor cells
OTX2	A member of the bicoid family of homeodomain transcription factors
RDH2	Retinol dehydrogenase 5—a component of the visual cycle
LCA3[SPAT7]	Spermatogenesis associated protein 7—a protein expressed in the retina but not in photoreceptor cilia
GUCY2D	Guanylate cyclase—another component of visual transduction
AIPL	Aryl hydrocarbon-interacting receptor protein-like 1—a protein which may have nuclear transport or chaperone activity
CRX	A photoreceptor transcription factor

References

1. http://www.who.int/blindness/cuases/en/
2. Ammann F, Klein D, Franceschetti A (1965) Genetic and epidemiological investigations on pigmentary degeneration of the retina and allied disorders in Switzerland. J Neurol Sci 2(2):183–196
3. Boughman JA, Conneally PM, Nance WE (1980) Population genetic studies on retinitis pigmentosa. Am J Hum Genet 32:223–235
4. Jay M (1982) On the heredity of retinitis pigmentosa. Br J of Ophthalmol 66:405–416
5. Puech B (1994) Prevalence of inherited retinal dystrophies in EUROPE, vol 3. Vilnius State Publishing Centre, Vilnius, pp 124–130
6. Bunker CH, Berson EL, Bromley WC, Hayes RP, Roderick TH (1984) Prevalence of retinitis pigmentosa in Maine. Am J Ophthalmol 97(3):357–365
7. Allikmets R (2004) Leber congenital amaurosis: a genetic paradigm. Ophthalmic Genet 25(2):67–79
8. Humphries P, Kenna P, Farrar GJ (1992) On the molecular genetics of retinitis pigmentosa. Science 256:804–808
9. Kennan A, Aherne A, Humphries P (2005) Light in retinitis pigmentosa. Trends Genet 21(2):103–110
10. Bhattacharya SS, Wright AF, Clayton JF, Price WH, Phillips CI, McKeown CM, Jay M, Bird AC, Pearson PL, Southern EM et al (1984) Close genetic linkage between X-linked retinitis pigmentosa and a restriction fragment length polymorphism identified by recombinant DNA probe L1.28. Nature 309(5965):253–255

11. McWilliam P, Farrar GJ, Kenna P, Bradley DG, Humphries MM, Sharp EM, McConnell DJ, Lawler M, Sheils D, Ryan C et al (1989) Autosomal dominant retinitis pigmentosa (ADRP): localization of an ADRP gene to the long arm of chromosome 3. Genomics 5(3):619–622

12. Farrar GJ, McWilliam P, Bradley DG, Kenna P, Sharp EM, Humphries MM, Lawler M, Eiberg H, Conneally MP, Trofatter JA, Humphries P (1990) Autosomal dominant retinitis pigmentosa: linkage to rhodopsin and evidence for genetic heterogeneity. Genomics 8:35–40

13. Dryja TP, McGee TL, Reichel E, Hahn LB, Cowley GS, Yandell DW, Sandberg MA, Berson EL (1990) A point mutation of the rhodopsin gene in one form of retinitis pigmentosa. Nature 343(6256):364–366

14. Farrar GJ, Kenna P, Redmond R, McWilliam P, Bradley DG, Humphries MM, Sharp EM, Inglehearn C, Bashir R, Jay M, Watty A, Ludwig M, Schinzel A, Sammans C, Gal A, Bhattacharya S, Humphries P (1991) Autosomal dominant retinitis pigmentosa (RP4): absence of the rhodopsin codon 23 proline->histidine transversion in pedigrees of European origin. Am J Hum Genet 47:941–945

15. Farrar GJ, Findlay JBC, Kumar-Singh R, Kenna P, Humphries MM, Sharp E, Humphries P (1993) Autosomal dominant retinitis pigmentosa: a novel mutation in the rhodopsin gene in the original 3q linked family. Hum Mol Genet 9:769–771

16. Farrar GJ, Kenna P, Jordan S, Kumar-Singh R, Humphries MM, Sharp EM, Sheils D, Humphries P (1991) A 3 base-pair deletion in the peripherin gene in one form of retinitis pigmentosa. Nature 354:478–480

17. Jordan SA, Farrar GJ, Kenna P, Humphries MM, Sheils D, Kumar-Singh R, Sharp EM, Soriano N, Ayuso C, Benitez J, Humphries P (1993) Localisation of an autosomal dominant retinitis pigmentosa gene to chromosome 7q. Nat Genet 4:54–57

18. Kennan A, Aherne A, Palfi A, Humphries MM, Stitt A, Simpson D, Demtroder K, Orntoft T. Ayuso C, Kenna PF, Farrar GJ, Humphries P (2002) Identification of an IMPDH1 mutation in autosomal dominant retinitis pigmentosa (RP10) revealed following comparative microarray analysis of transcripts derived from retinas of wild-type and Rho-\- mice. Human Mol Genet 11(5):547–558

19. Farrar GJ, Kenna P, Humphries P (2002) Molecular genetics of Retinitis Pigmentosa: mutation independent approaches to therapy. EMBO J 21(5):857–864

20. https://sph.uth.tmc.edu/retnet/

21. D'Cruz PM, Yasumura D, Weir J, Matthes MT, Abderrahim H, LaVail MM, Vollrath D (2000) Mutation of the receptor tyrosine kinase gene Mertk in the retinal dystrophic RCS rat. Hum Mol Genet 9(4):645–651

22. Sung CH, Makino C, Baylor D, Nathans J (1994) A rhodopsin gene mutation responsible for autosomal dominant retinitis pigmentosa results in a protein that is defective in localization to the photoreceptor outer segment. J Neurosci 14(10):5818–5833

23. Travis GH (1998) Human Genetics'98: Apoptosis. Mechanisms of cell death in the inherited retinal degenerations. Am J Hum Genet 62:503–508

24. Mendes H, Cheetham M (2008) Pharmacological manipulation of gain-of-function and dominant-negative mechanisms in rhodopsin retinitis pigmentosa. Hum Mol Gen 17(19):3043–3054

25. Vithana EN, Abu-Safieh L, Allen MJ, Carey A, Papaioannou M, Chakarova C, Al-Maghtheh M, Ebenezer ND, Willis C, Moore AT, Bird AC, Hunt DM, Bhattacharya SS (2001) A human homolog of yeast pre-mRNA splicing gene, PRP31, underlies autosomal dominant retinitis pigmentosa on chromosome 19q13.4 (RP11). Mol Cell 2:375–381

26. McKie AB, McHale JC, Keen TJ, Tarttelin EE, Goliath R, van Lith-Verhoeven JJ, Greenberg J, Ramesar RS, Hoyng CB, Cremers FP, Mackey DA, Bhattacharya SS, Bird AC, Markham AF, Inglehearn CF (2001) Mutations in the pre-mRNA splicing factor gene PRPC8 in autosomal dominant retinitis pigmentosa (RP13). Hum Mol Genet 10(15):1555–1562

27. Chakarova CF, Hims MM, Bolz H, Abu-Safieh L, Patel RJ, Papaioannou MG, Inglehearn CF, Keen TJ, Willis C, Moore AT, Rosenberg T, Webster AR, Bird AC, Gal A, Hunt D, Vithana EN, Bhattacharya SS (2002) Mutations in HPRP3, a third member of pre-mRNA

splicing factor genes, implicated in autosomal dominant retinitis pigmentosa. Hum Mol Genet 11:87–92

28. Keen TJ, Hims MM, McKie AB, Moore AT, Doran RM, Mackey DA, Mansfield DC, Mueller RF, Bhattacharya SS, Bird AC, Markham AF, Inglehearn CF (2002) Mutations in a protein target of the Pim-1 kinase associated with the RP9 form of autosomal dominant retinitis pigmentosa. Eur J Hum Genet 10(4):245–249

29. Mordes D, Luo X, Kar A, Kuo D, Xu L, Fushimi K, Yu G, Sternberg P, Wu JY (2006) Pre-mRNA splicing and retinitis pigmetosa. Mol Vis 12:1259–1271

30. Zhao C, Bellur DL, Lu S, Zhao F, Grassi MA, Bowne SJ, Sullivan LS, Daiger SP, Chen LJ, Pang CP, Zhao K, Staley JP, Larsson C (2009) Autosomal-dominant retinitis pigmentosa caused by a mutation in SNRNP200, a gene required for unwinding of U4/U6 snRNAs. Am J Hum Genet 85:617–627

31. Tanackovic G, Ransijn A, Ayuso C, Harper S, Berson EL, Rivolta C (2011) A missense mutation in PRPF6 causes impairment of pre-mRNA splicing and autosomal-dominant retinitis pigmentosa. Am J Hum Genet 88(5):643–649

Chapter 2
Gene-Based Medicines Targeting Genetic Defects Directly and Molecular Pathologies Common to Multiple Forms of Disease

Leber congenital amaurosis (LCA) was an attractive initial target for gene-based therapies targeting retinopathies for several reasons. Firstly, the disease is transmitted in an autosomal recessive mode, and hence gene replacement therapy, as opposed to a requirement to selectively target dominant acting mutations either at the DNA or transcript levels, would be required. Secondly, while, in many forms of disease congenital blindness is evident, photoreceptors remain viable, though non-functional. Since some forms of LCA are caused by defects in enzymes expressed in the pigmented epithelium, which re-cycle all trans retinal into the chromophore 11-cis retinal, it was envisaged that restoration of gene function within the RPE could well result in an immediate improvement of vision, in other words, a rapid and effective readout of treatment efficacy. A major success in gene therapy was reported by Acland et al. [1]. These workers targeted the Briard dog, a naturally occurring model of LCA. These animals have a four base pair deletion in the RPE65 gene and animals homozygous for this deletion are congenitally blind. Acland et al., injected sub-retinally into these animals an adeno-associated virus (AAV2/5) expressing a functional RPE65 gene driven by a chicken ß-actin promoter. 150–200 µl of AAV (at a titre of 2.3×10^{11} infectious particles per ml), was injected sub-retinally into the eyes of dogs. Post-treatment, animals were shown to regain retinal function, assessed by a substantial increase in rod derived electroretinographic (ERG) waveforms. These encouraging results have now lead to the initiation of clinical trials in human subjects, a total of nine such trials for LCA having so far been registered (Clinicaltrials.gov) in the US, the UK, France and Israel, all of these studies employing AAV vectors for gene delivery.

Several gene therapy trials for retinitis pigmentosa have also been registered targeting diseases induced by recessive mutations within the Mer thymidine kinase (MERTK) and MyosinVIIA genes. In regard to MERTK, the protein encoded by this gene is a receptor tyrosine kinase, homozygous mutations within which were initially shown by D'Cruz et al. [2] to cause a degenerative retinopathy in the Royal College of Surgeons (RCS) rat. Subsequently, mutations within this gene were found in some patients with autosomal recessive RP by Andreas Gal and

P. Humphries et al., *Hereditary Retinopathies*, SpringerBriefs in Genetics, DOI: 10.1007/978-1-4614-4499-2_2, © The Author(s) 2012

colleagues [3]. Successful gene replacement therapy in the RCS rat, using an AAV virus expressing the MERTK gene, was reported by Smith and colleagues in 2003 [4] and in 2005, Tschernutter et al. reported long-term preservation of the function of the retina in the RCS rat using a lentiviral vector [5]. A human clinical trial involving gene replacement using an AAV2 vector is currently in phase 1 in Saudi Arabia (Clinicaltrials.gov NCT01482195).

Mutations within the myosin VIIA gene were shown initially by Gibson et al. [6] to underlie the autosomal recessive disease manifested in the *Shaker 1* mouse, these animals displaying symptoms of vestibular (balance) dysfunction. In parallel, Weil et al. [7] discovered mutations within the same gene in the human syndrome, Usher syndrome type 1B. This autosomal recessive condition involves progressive auditory impairment together with retinitis pigmentosa. Hashimoto et al. [8] reported that lentiviral vectors could accommodate cDNAs large enough to encode Myosin XIIa and demonstrated rescue of the retinas of MYO7A null mice. A phase I/II clinical trial has been registered (Clinicatrials.gov NCT01505062) in the UK in which a medication, *UshStat,* will be used. The latter is a product of the company, Oxford Biomedica, and is a lentiviral vector expressing the Myosin VII gene. The same company has also registered a clinical trial of their product, StarGen, for treatment of Stargardt macular dystrophy. This is a rare autosomal recessive retinopathy involving degeneration of the cone-rich central portion of the retina, the macula. A major feature of this disease is a progressive buildup within the retinal pigmented epithelium of lipofuscin, a major component of which is N-retinylidene-N-retinyl ethanolamine (A2E), which is regarded as a major pathological feature [9]. Stargardt disease is caused by mutations within the ABCA4 gene, which were first identified by Rando Allikmets and colleagues [10], the gene encoding a transporter located in the outer segments of rod and cone photoreceptors, which assists in the transport of vitamin A intermediates from the photoreceptors into the retinal pigmented epithelium. Kong and colleagues subsequently showed that a lentiviral vector expressing the ABCA4 gene from a constitutive CMV or Rhodopsin promoter, subretinally introduced into the retinas of ABCA4 mice significantly reduced the accumulation of A2E and suggested that lentiviral gene therapy would likely represent an efficient method of treating this disease [11]. The StarGen product is a lentiviral vector expressing the ABCA4 gene.

A gene therapy trial has also been registered in the UK (Clinicaltrials.gov NCT01461213) in which an AAV vector will be used in phase I/II trials for treatment of choroideremia. The latter is a rare X-linked recessive condition in which the choroid underlying the retina degenerates and is caused by mutations within the RAB Escort protein 1 (REP-1). In this case, the medication is an AAV2 vector expressing the REP-1 gene.

Over the last 10 years, a growing number of proof of efficacy studies have been reported in which gene replacement strategies have been used in rodent and other animal models of retinopathy and many of these undoubtedly represent the forerunners of human gene therapy trials. The majority of these have involved the use of AAV vectors. For example, Pawlyk et al. [12] reported successful gene replacement therapy in a mouse model of LCA caused by null mutations within the

RPGRIP gene, demonstrating correct localization of the protein to the photoreceptor cilia, and preservation of function of photoreceptors by electroretinography. Additional studies validating gene therapy for forms of LCA caused by mutations in other genes were reported in 2010 and 2011 (see below).

Successful gene replacement therapy was also reported in 2005 by Min et al. [13], in a mouse model of retinoschisis. This is a rare X-linked retinopathy characterized by loss of central vision and with the appearance of small cysts within the retina. The retina can also split peripherally, between the inner and outer nuclear layers (hence the name of the disease). The disease is caused in males by a null mutation within the retinoschisin gene (Rs1), an AAV vector expressing RS1 cDNA having been shown by these workers to restore retinoschisin expression and to preserve retinal function and morphology. Additional studies on gene replacement therapy for retinoschisis were reported by Kjellstrom and colleagues [14]. These workers injected an AAV2/2 vector subretinally into Rs1h knockout mice at post-natal day 14 and examined retinal structure and function at 14 months, where improved outer and inner retinal structure were observed together with improved ERG amplitudes. These studies were undertaken using the subretinal route of AAV delivery, which induces a detachment of the retina in the region of the inoculations, which subsequently reattaches. Park and colleagues [15] reasoned that an intravitreal route of viral administration might be more feasible in a human scenario in that the procedure does not detach the retina, which is innately fragile in retinoschisis. They were able to demonstrate in retionschisin knockout mice efficient transduction of the retina following intravitreal injection of the virus.

Further evidence of the efficacy of AAV-mediated gene replacement therapy for retinal degenerations was provided by Alexander et al. [16] in a study targeting achromatopsia. This is an exceedingly rare autosomal recessive retinopathy affecting no more than 1 in 30,000 people in which there is loss of color vision. Defects within four different genes have been identified in different individuals or families. These encode three of the subunits of cyclic nucleotide gated channels, CNGA3, CNGB3 (the alpha and ß subunits of cone photoreceptor cGMP-gated channel respectively), GNAT2 (guanine nucleotide binding protein 2), in addition to the alpha subunit of cone cGMP phosphodiesterase (PDE6C). Mutations within these genes are encountered with widely differing frequencies, the most common being CNGB3 in about half of all cases of the disease. These workers demonstrated rescue of cone-ERG responses and visual acuity in a murine model of this disease. In a later study, Komaromy and colleagues [17] used an AAV5 vector expressing the CNGB3 gene in a naturally occurring dog model of achromatopsia and using a red opsin promoter they were able to show a therapeutic effect for up to 3 years post inoculation. Quoting the authors: 'Our results hold promise for future clinical trials of cone-directed gene therapy in achromatopsia and other cone-specific disorders'. More recently in 2011, Carvalho and colleagues used an AAV2/8 vector expressing a human cone-arrestin-driven human CNGB3 gene subretinally inoculated in *CNGB3−/−* mice and were able to demonstrate highly significant restoration of cone function—'This study represents achievement of the most substantial restoration of visual function reported to date in an animal model of achromatopsia using a human gene construct, which has the potential to be utilized in clinical trials' [18]. In an additional study of this disease, Michalakis and colleagues [19] reported that cone vision could be restored in *CNGA3–/–* mice using an AAV vector and in this study were able to demonstrate activation of the output neurons of the retina, ganglion cells.

Efficacy of AAV-mediated gene replacement was also demonstrated in 2008 in the naturally occurring rd10 mouse model of autosomal recessive RP. These animals are homozygous for null mutations within the gene encoding the ß-subunit of rod photoreceptor cGMP phosphodiesterase, a central enzymatic component of the visual transduction cycle. Pang and colleagues injected an AAV5 vector expressing the ßPDE gene subretinally into the animals at P14 and examined animals after a further period of 3 weeks, both scotopic and photopic ERG waveforms being notably preserved, as well as good preservation of the outer nuclear layer of the retina and of photoreceptor outer segments [20]. This work was extended in 2011 with a demonstration that preservation of vision was achievable for at least 6 months post AAV8 injection [21].

While gene therapy trials are well established for LCA involving null mutations within the RPE65 gene, mutations within 16 genes, as outlined above, have been implicated in this condition. One of these is the gene encoding aryl hydrocarbon receptor-interacting protein-like 1 (AIPL1). Mutations within this gene not only cause LCA, but have been encountered in some forms of RP and cone-rod dystrophy. Tan and colleagues [22], reported the use of AAV2/2 and AAV2/8 vectors as gene replacement vehicles in several murine models of LCA bearing mutations within this gene and were able to demonstrate significant rescue of the photoreceptor degeneration. In a subsequent study, Sun and colleagues [23] were also able to demonstrate significant and long-term retinal rescue in mice expressing mutated AIPL1 genes using an AAV vector. Pawlyk and colleagues [24] also investigated gene replacement therapy for LCA in a murine model of disease caused by null mutations within the RPGRIP1 gene, this time using an AAV8 vector expressing a human gene replacement driven by a rhodospin kinase promoter. Significant therapeutic benefit was obtained, the authors commenting: 'This study demonstrates the efficacy of human gene replacement therapy and validates a gene therapy design for future clinical trials in patients afflicted with this condition'. Another gene involved in LCA (LCA1) encodes guanylate cyclase 1 (Gucy2d), mutations within this gene being estimated to cause up to 20 % of cases of LCA. Boye and colleagues have shown that subretinal inoculation of an AAV5 vector expressing the murine GC1 gene driven either by a rhodopsin kinase or a ubiquitous chicken ß-actin promoter resulted in extensive restoration of cone function in treated animals. Quoting the authors: 'These results lay the ground work for the development of an AAV-based gene therapy vector for the treatment of LCA1' [25]. A further study involving AAV-mediated gene replacement of guanylate cyclase 1 in a mouse LCA1 model was reported by Mihelec and colleages [26]. They subretinally inoculated an AAV2.8 vector expressing a human rhodopsin kinase promoter-driven human GUCY2D gene. The authors concluded: '…we observed a dose-dependent restoration of rod and cone function and an improvement in visual behavior of transduced area up to 6 months post injection. To date, this is the most effective rescue of the Gucy2e−/− mouse model of LCA and we propose that a vector, similar to the one used in this study, could be suitable for use in a clinical trial of gene therapy for LCA1'.

Gene therapies have also been validated in murine models of a number of hereditary syndromes that incorporate RP. AAV-mediated gene replacement therapy has been validated in a murine model of Usher syndrome type 2D, with a targeted disruption of the DFNB31 gene, encoding the PZD scaffold protein, Whirlin [27]. In that study, Zou and colleagues [27] subretinally inoculated an AAV2/5 vector into DFNB31−/− mice expressing a rhodopsin kinase promoter driven replacement gene. They comment: 'The combined hRK [kinase promoter] and AAV gene delivery system could be an effective gene therapy approach to treat retinal degeneration in USH2D patients'. A syndromic form of RP involving microthalmos, foveoschisis and drusen deposition at the optic disk, is caused by mutations within the membrane-tuppe frizzled-related protein (MFRP). In another example of AAV-mediated retinal gene therapy, Dinculescu and colleagues have recently shown in a mouse model of retinal degeneration with mutations in this gene (the rd6 mouse) that an AAV8 vector expressing the murine gene driven by a chicken ß-actin promoter subretinally injected at day 14 rescues rod and cone photoreceptors as assayed at 2 months of age [28]. Another well known, though rare syndrome incorporating RP is Bardet-Biedl syndrome. This (usually recessive) condition may feature hypogonadism, mental retardation, polydactyly, obesity and kidney disease and displays extensive genetic heterogeneity, to date, mutations in 15 different genes having been implicated in the disease, which, when mutated result in ciliary dysfunction. One of the consequences of this is that the primary photoreactive pigment of rod cells, rhodopsin, cannot be transported from the inner to the outer segments of the photoreceptors and photoreceptor cell death occurs as a result. The Bbs4 null mouse is a naturally occurring model of Bardet-Biedle syndrome. Simons and colleagues [29] have recently shown that subretinal inoculation of an AAV5 vector expressing the mouse Bbs4 gene results in correct localization of rhodopsin and preservation of rod cells. The authors conclude: '…our treatment prevents photoreceptor death, improves electrophysiological function of the retina and preserves visually evoked behavioral responses. These data from a mammalian model suggest that BBS-associated retinal degeneration can be treated'.

It is notable that all human clinical trails so far registered, and all animal studies outlined above, have involved treatment of autosomal recessive or X-linked recessive forms of disease. Here, both versions of the gene in question are non-functional, and therefore a strategy of gene replacement can be employed. However, while essentially all forms of LCA are recessive, up to 30 % of cases of RP display autosomal dominant modes of transmission. Here, the gene product is either 'haplo-insufficient', perhaps a rather cumbersome term, simply meaning that one allele is nonfunctional and that expression from the remaining functional gene is insufficient to maintain photoreceptor viability, or alternatively, one gene homolog is functionally normal, while the mutated version exerts a 'dominant negative' effect in causing photoreceptor cell mortality. While the distinction between these two basic mechanisms of photoreceptor cell death remains to be fully established for all genes or mutations, many dominant forms of RP fall into the latter category, including, for example, those caused by mutations within the rhodopsin, RDS-peripherin, and IMPDH1 genes. A substantial number of studies in animal models of autosomal dominant retinal disease have been reported in which therapeutic efficacy has been obtained using methods that target disease-causing genes directly. Strategies have included RNAi-mediated down regulation of both normal and mutant alleles at the same time, which in some cases has been shown to be of therapeutic benefit because lack of expression of both alleles results in only a mild phenotype. Alternatively both normal and mutant alleles can be suppressed, followed by introduction of a functional gene that has been modified such as to escape RNAi-mediated suppression. Another strategy has involved overexpression of the normal allele such as to reduce, in relative terms, the amount of mutant gene product.

The concept of a gene-based mutation-independent approach to therapy for autosomal dominant disease was published initially by Millington-Ward and colleagues [30] and reviewed by Farrar et al. [31]. According to this concept, a gene encoding shRNA targeting transcripts derived from a dominant-acting gene is introduced into an AAV vector together with a replacement gene which has been structurally modified at third base degenerate codon sites such as to encode a normal transcript that can not be bound by the shRNA. In this way, a single shRNA molecule can be used to down regulate both normal and mutant transcripts, while at the same time a functional gene encoding transcripts resistant to the shRNA is introduced. This way a single gene therapy vector can, in principle, be used in targeting multiple mutations within a given gene. This is especially relevant, where multiple mutations have been identified within a given gene, such as is the case for rhodopsin, where up to 100 different mutations have so far been encountered in patients with autosomal dominant RP. Studies of this approach in a murine dominant RP model with a mutation in the rhodopsin gene were reported by O'Reilly et al. [32], Chadderton et al. [33], Palfi et al. [34] and Millington-Ward et al. [35]. In an alternative approach, Mao et al. [36] demonstrated that subretinal inoculation of an AAV vector expressing a normal murine rhodopsin gene into mice expressing a dominant acting mutation within the rhodopsin gene (P23H) induced significant protection of the photoreceptors by decreasing, in relative terms the amount of mutant rhodopsin present in rod cells. In an RP10 model of autosomal dominant RP, simultaneous ablation both of normal and dominantly mutated transcripts has been shown to have therapeutic benefit. This form of disease is caused by dominant acting mutations within the inosine monophosphate dehydrogenase gene (IMPDH1). In humans the disease is early in onset, producing significant visual handicap within the second decade of life, current evidence indicating that disease pathology is induced by aggregation of mutant IMPDH1 protein [37]. On the other hand, mice with a targeted disruption of the IMPDH1 gene display a much slower retinal degeneration and even at 1 year of age (equivalent to late middle age in humans) retain much of their outer nuclear layer and retinal function [37]. Tam and colleagues [38] have shown that subretinal inoculation into mice of an AAV virus expressing shRNA directed towards a mutant human IMPDH1 together with the equivalent endogenous murine transcript, substantially alleviates disease progression.

Given the North Face of genetic heterogeneity that must be surmounted in developing gene-based medicines targeting individual genes, there is clearly a strong rationale in looking for gene-based products that are not required to target primary genetic defects, but rather, molecular pathologies common to more than one, and hopefully many individual retinal conditions. Significant progress is being made in this area and has included studies of the efficacies of retinal protection afforded by expression within the retina of rod-derived cone viability factor (RDCVF), glial-derived neurotrophic factor (GDNF), ciliary neurotrophic factor (CNTF), brain-derived neurotrophic factor (BDNF), the chaperone, Grp78/BiP, lens epithelium-derived growth factor, C1q, X-linked inhibitors of apoptosis (XIAP) and of genes encoding artificial light sensors, including channelrhodopsin and halorhodopsin in output neurons of the retina (ganglion cells) or indeed in surviving, but nonfunctional cone cells, this approach having been referred to as optogenetic therapy.

It is of note that the primary degenerative feature of retinitis pigmentosa is the progressive death of rod photoreceptors, resulting in the initial symptom of the disease—nyctalopia. However, cone photoreceptors persist for longer periods although they can remain viable for substantial periods of time after their functional activity has been lost. In addition, the output ganglion cells of the retina persist long after photoreceptors have degenerated. Optogenetics refers to the concept of genetically modifying ganglion cells or surviving nonfunctional cones by introducing into them genes that encode membrane channels that can be directly activated by light. Two such channels have been extensively explored, channelrhodopsin-2 from *Chlamydomonas reinhardtii*, a flagellated green algae, and halorhodopsin, originally identified in the halophylic microorganism, *Natronomonas pharaonis*, originally isolated from a salt lake in Africa. When inserted into the membranes of neurons, light activation of channelrhodopsin-2 results in direct cell depolarization. On the other hand, light activation of halorhodopsin results in cell hyperpolarization. Bi et al. [39] incorporated a gene encoding channelrhodopsin-2 and green fluorescent protein into an AAV2 vector driven by a hybrid chicken ß-actin-CMV promoter. Intravitreal injection of this virus into rd1−/− mice, in which photoreceptor cells were absent, resulted in transfection of retinal ganglion cells and extensive expression within the inner plexiform layer of the retina. It is of note that stable transfection of the retina was achievable for at least 12 months. Dissociated retinal ganglion cells expressing channelrhodopsin-2 could be identified by virtue of green fluorescence, patch clamp recordings from individual cells indicating light-evoked responses were generated and in addition, visual evoked responses were recordable in such animals. However, the authors pointed out that AAV-mediated transfection of ganglion cells targets both ON- and OFF-ganglion cells at the same time and how this would impact on visual perception remained to be established. Restoration of visual response by channelrhodospin in the RCS rat has also been reported by Tomita and colleagues [40, 41]. Busskamp and colleagues [42] have more recently incorporated a halorhodopsin gene driven either by a red cone opsin or cone arrestin-3 promoter into an AAV vector, with the objective of transfecting viable, though nonfunctional cone photoreceptors in slow and fast mouse models of retinal degeneration (*CNGA3−/−Rho−/−* and *rd1* respectively). The advantage of halorhodopsin is that in response to light-activation, transfected neurons are hyperpolarized, as is the case in natural light activation of cone photoreceptors. Visual evoked potentials could be detected in treated animals and in both dark light box and optimotor tests, treated animals performed better. Moreover, these authors have identified patients with RP whom, by OCT analysis have no photoreceptor outer segments but who retain cone cell bodies and potentially these individuals could be subjects for clinical trials.

Extensive studies have been undertaken into the efficacy of expression neurotrophic factors in protecting photoreceptors. Faktorovich and colleagues, in an early pioneering study reported in 1990, that intravitreal inoculation of basic fibroblast growth factor (ßFGF) directly into the vitreous humor of RCS rats, a model of autosomal recessive RP homozygous for null mutations within the MERTK gene, afforded extensive protection to photoreceptor cells [43]. LaVail and colleagues demonstrated in 1992 that direct intravitreal inoculation of a variety of neurotrophic factors into the eyes of wild type rats, protected the retinas from the effects of light damage [44]. Effective agents included bFGF, BDNF, CNTF, interleukin 1ß, and acidic fibroblast growth factor. Subsequently, La Vail and colleagues also reported results of direct intravitreal inoculation of a variety of neurotrophic factors into mouse and rat retinopathy models and concluded that retinal degeneration can be slowed down by CNTF and leukemia inhibitory factor and in some cases by BDNF [45]. Given the proinflammatory role of IL-1ß, Whiteley and colleagues [46], also demonstrated that doses of IL-1ß much lower than had previously been used, still afforded retinal protection while minimizing inflammatory damage in the RCS rat model. The authors conclude, 'Since the survival-promoting effects of IL-1ß are comparable to those of bFGF, even when given at much lower dosages, this cytokine may play an important role in the phenomenon of injury-related PRC [photoreceptor cell] rescue and represents another potentially useful survival factor for retinal degenerative diseases' [46]. Cayouette and colleagues, subsequently, demonstrated that intravitreal injection of an adenoviral vector expressing CNTF significantly protected the retinas of rd−/− mice, with a naturally occurring null mutation of the gene encoding the ß subunit of cGMP phosphodiesterase [47]. They also found that intravitreal inoculation of CNTF produced a protective effect on the retina, but less so than expression from the AV genome. Liang and colleagues [48] were also able to demonstrate a protective effect of this polypeptide in Rho−/− mice. These authors incorporated a CNTF gene along with a GFP gene into an AAV vector, which was subretinally inoculated into the rhodopsin knockout mice. In such animals, all rod photoreceptors die over a period of approximately 3 months, and in these experiments, expression of CNTF produced a protective effect over the entire 3-month period.

An additional study was reported by these authors in 2001 in which an AAV vector expressing CNTF was intravitreally as well as subretinally inoculated into the eyes of rds−/− mice and into the eyes of P23H and S334ter rhodopsin transgenic rats [49]. They found significant structural protection of the photoreceptors in all animals examined, although in treated rats, ERG amplitudes were lower than controls. The authors suggest: 'the need for greater understanding of the mechanism of action of CNTF before human application can be considered'. Further studies of the potentially beneficial effects of CNTF therapy were reported in 2002 by Bok and colleagues [50]. These authors introduced AAV vectors expressing a CNTF gene driven either by CMV or chicken ß-actin promoters into the retinas of mice expressing a dominant mutation within the rds-peripherin gene. They observed long-term pan-retinal protection of the retinas of these animals, which was more readily achievable when the virus was subretinally inoculated. However, as noted previously in other models, both scotopic a and b wave ERG amplitudes were reduced in treated animals. The authors conclude: '…proper doses of CNTF administration should be determined before human clinical trials are considered for the amelioration of inherited retinal degenerations with CNTF'. Schlichtenbrede and colleagues [51], using rds−/− mice, while observing improved photoreceptor structure as a result of intravitreal injection of AAV expressing CNTF, also saw a marked deterioration in retinal function assessed by ERG, and even found such deterioration in wild type mice similarly injected. They conclude: 'Our results demonstrate that intraocular CNTF gene delivery may have a deleterious effect on the retina and caution against its application in clinical trials'. Adamus and colleagues [52] studied the effects of AAV-mediated expression of CNTF in a mouse model of cancer-associated retinopathy. In humans, this is a rare syndrome and when it occurs it is most frequently associated with lung carcinoma. The disease is associated with the presence of autoantibodies against the retina expressed protein, recoverin, and can be modeled in mice by inoculating anti-recoverin antibodies. These researchers subretinally inoculated an AAV vector expressing CNTF into mice that were then inoculated with anti-recoverin antibody to induce a retinal degeneration. The authors found a significant protective effect in retinas expressing CNTF and conclude: 'CNTF may be a useful treatment for human antibody-mediated retinal degeneration'. In an additional study reported in 2004, Huang and colleagues examined the effect of direct inoculation into the eye of recombinant CNTF as well as AV-mediated expression of the cytokine in the RCS rat [53]. They found that photoreceptor cell death was potently delayed and that treated animals showed improved retinal function assessed by ERG. They conclude: 'this study indicates that adenoviral CNTF effectively rescues degenerating photoreceptors in RCS rats'. In 2006, Buch and colleagues reported deleterious effects of AAV-mediated expression of CNTF were dose dependent, but administering levels of CNTF that did not affect photoreceptor function, did not protect photoreceptors from degeneration [54]. In the same study, however, they reported that AAV-mediated expression of GDNF afforded protection to photoreceptors in two rodent models of RP. In humans, a phase 1, and two phase II clinical studies have been undertaken using Encapsulated Cell Technology (Neurotech Inc) in which a small 6 mm long implant containing human cells genetically engineered to secrete CNTF is surgically implanted in a removable way into the vitreous of the eye. In a recent study, cone cell structure and function were reported in detail on three patients having received such implants, one patient with autosomal dominant RP with a rhodopsin mutation, one with Usher syndrome type 2 and one with simplex RP [55]. The authors conclude: 'In this study presenting the first images of cone photoreceptors in human eyes treated with CNTF, the results suggest that CNTF may slow cone photoreceptor loss in eyes with retinal degeneration. They also provide evidence to support the pursuit of additional, larger prospective, masked clinical trials of CNTF using AOSLO images as an outcome measure of disease progression and treatment response'.

Many studies of the protective effects of GDNF on photoreceptors have been reported. Frasson and colleagues [56] investigated the effects of direct intraocular injection of GDNF in the *rd−/−* mouse, a naturally occurring model of retinal degeneration homozygous for null mutations within the RDS-peripherin gene. While at 22 days of age rod-derived ERG responses were unrecordable in untreated animals, a proportion of treated animals displayed improved retinal function and all animals treated had significantly greater numbers of rod cells. Given that mutations within the RDS-peripherin gene are found in human RP, these authors concluded, 'GDNF represents a candidate neurotrophic factor for palliating some forms of hereditary human blindness'. McGee and colleagues [57] investigated the protective effects of AAV-mediated expression of GDNF in photoreceptor cells of a rat model of RP expressing a dominant S334-ter mutation within the rhodopsin gene. While a significant protection of the retina was observed, the effects diminished after 60 days as the photoreceptors expressing the cytokine degenerated. Wu and colleagues [58] examined the long-term safety of delivering a gene expressing GDNF incorporated into an AAV vector by the intravitreal route in wild type Sprague–Dawley rats. Accumulation of GDNF was immunologically confirmed for a period of up to 1 year, and there appeared to be no negative effects on retinal function or histology. In 2006, Hauck and colleagues showed that receptors for GDNF, GFRalpha, and RET are on Muller glial cells but not on photoreceptors in pig retinas [59]. They showed that GDNF induces ERK phosphorylation and that such signaling results in the induction of FGF-2. Thus, the protective effects of GDNF were shown to be mediated by Muller cells. In 2007, Dong and colleagues reported a study in which degeneration of the retinas of mice was induced by oxidative damage following treatment with paraquat, $FeSO_4$, and hyperoxia [60]. These were transgenic animals incorporating a GDNF gene that was inducible within the retina by doxycycline. They observed a significant positive effect on retinal function as assessed by ERG and on retinal morphology. These authors concluded that their data 'suggest that gene transfer of GDNF should be considered as a component of therapy for retinal degenerations in which oxidative damage plays a role', therefore suggesting that this neurotrophic factor may prove to be of value in treatment of age-related macular degeneration in addition to hereditary retinopathies such as RP. Further indication of the benefits of GDNF expression within the retina was provided by Gregory-Evans and colleagues [61]. They used mouse embryonic stem cells generated to express elevated levels of GDNF and up to 200,000 of such cells were intra-vitreally inoculated into the retinas of rats expressing a dominant-acting mutation (S334ter) within the rhodopsin gene—a model of autosomal dominant RP at day 21 post-birth. Stem cells expressing GDNF were still detectable in the retinas of these animals at day 90, photoreceptor cell counts being significantly higher in treated animals. Further studies on the mechanisms of neuroprotection induced by GDNF were reported in 2011 by Del Rio and colleagues, in a study of transcripts induced in retinal tissues by GDNF [62]. One of these encoded the cytokine, osteopontin (OPN). OPN was further shown to display a protective effect on photoreceptors in culture and on retinal explants from *Pde6brd1* mice, the latter an autosomal recessive retinopathy, the authors concluding that these findings 'suggest that RMG [retina muller glia]-derived OPN is a novel candidate protein that transmits part of the GDNF-induced neuroprotective activity of RMG to PR [photoreceptor] cells'. In a recent (2011) study, Dalkara and colleagues [63] targeted expression of GDNF to retina Muller glial cells, reasoning that these cells are natural secretors of neurotrophins, that they span essentially the entire retina and that they persist even in extensively degenerating retinas. The AAV used in these experiments was a capsid variant that was able to selectively transduce Muller cells [64]. Expression of GDNF from Muller cells leads to long-term protection of the retinas of transgenic rats expressing a dominant-acting mutation (Ser334-ter) in the rhodopsin gene, a model of RP.

It has been established for some time that introduction of wild type murine photoreceptors into the eyes of *rd−/−* mice protected cone photoreceptors from degeneration and that the protective effect appeared to be the result of a diffusible factor which was probably a protein [65–68]. In order to identify the gene or genes involved, Leveillard and colleagues screened up to 200,000 cDNAs derived from a normal mouse retina in an in vitro cone cell survival assay and identified a clone, encoding a protein which was termed RdCVF [69]. The RdCVF gene (Nxnl1) encodes a thiol-oxidoreductase enzyme and is believed to protect the retina from oxidative stress [70]. Supporting the protective role of this protein, *Nxnl1−/−* mice display a gradual degeneration in cone function and injection of the purified protein into the retinas of transgenic rats expressing the common P23H mutation within the rhodopsin gene resulted in significant protection of cone photoreceptors [71]. These data suggest that retinal protection could be achieved in genetically diverse forms of retinopathy by introduction into the retina of an AAV vector expressing the NXNL1 gene to compensate for the loss of this secreted protein as rod cells die off.

A number of approaches to retinal protection have been explored which involve the direct targeting of apoptotic processes, approaches involving both inhibition of apoptosis, and enhancement of its efficiency, having been explored. In regard to the former, the rationale was to preserve photoreceptors by suppressing the mechanism inducing their death. In regard to the latter, the reasoning was that if apoptotic mechanisms were enhanced, dying photoreceptors may be engulfed more efficiently by the RPE or by resident retinal macrophages, thus reducing the number of cells dying by necrotic processes, where the contents of cells would be liberated, thus exacerbating inflammatory processes which may be to the detriment of remaining retinal neurons. Apoptosis, or programmed cell death, involves the condensation of dying cells into apoptotic bodies, which are readily phagocytosed, and many studies have indicated that it is a common mechanism of cell death in degenerative retinopathies [72–74]. A number of studies in which the ability of anti-apoptotic members of the Bcl family, Bcl-2 and the related protein Bcl-XL, to slow down the death of photoreceptors in animal models of degenerative reti-nopathy have been reported [75–78]. Bcl-2 exerts its protective effect by inhibiting the export of cytochrome C from photoreceptor mitochondria [79, 80]. However, somewhat mixed results have been obtained from these experiments. Chen et al. [76] generated transgenic mice in which relatively high levels of Bcl-2 protein were expressed using an opsin promoter to drive the gene. They crossed these animals with mice expressing a truncated rhodopsin gene (S334ter) in which retinal degeneration is normally rapid, all photoreceptors degenerating in such animals by 3 weeks of age. They also crossed Bcl-2 expressing mice onto an rd−/− background, rd−/− mice normally losing their photoreceptors, again within about 3 weeks of birth. The effects of elevated expression of Bcl-2 on light-induced photoreceptor degeneration, where, after 14 days of constant light expo-sure, most photoreceptors are normally dead, were also tested. In all three sce-narios, significant protection of photoreceptors was achieved. The authors concluded, 'These results strongly support the idea that retinal degeneration may be treated by modulating the entry of PRs [photoreceptors] into the apoptotic cell death pathway'. Tsang and colleagues [77] undertook similar experiments in which mice with a homozygous mutation within the gamma subunit of retinal cGMP phosphodiesterase, normally developing a very rapid retinal degeneration where photoreceptors are essentially all dead within 4 weeks of birth, were crossed with a strain of mouse expressing an opsin-driven Bcl-2 gene. The results of these experiments clearly indicated a protective effect induced by expression of Bcl-2, but this was only of limited duration. Joseph and Li [75], however, were unable to demonstrate any protective effect of expression of Bcl-2, or Bcl-XL in the retinas of rd−/− mice and in mice expressing a dominant mutation (K296E) within the rhodopsin gene, nor were they able to show any protective effect of these genes against light damage. Nir and colleagues [78] investigated the possible protective effects of expression of Bcl2 in the retinas of rds−/− mice. These animals were homozygous for recessive null mutations within the rds-peripherin gene, mutations within which are known to cause both recessive and dominant forms of RP as well as forms of macular degeneration. At 3 months of age, the retinas of rds−/− mice normally have about three rows of photoreceptor nuclei in their outer nuclear layer. When crossed onto mice expressing a Bcl-2 gene driven by a mouse rho-dopsin promoter, up to six rows of photoreceptor nuclei were observed, the authors concluding that 'Bcl-2 expression may be an effective therapeutic strategy in humans with mutations in the RDS or other genes that affect the integrity of photoreceptor outer segments'. The reasons for such discrepancies in these studies are unclear, although rds−/− mice display a relatively slow degeneration in comparison to the other models of retinal degeneration that were employed, which may partially account for the differences in protective effects observed. To date, no clinical trial involving expression of Bcl-2 or Bcl-XL has been registered.

Given the commonality of apoptosis as a mechanism of photoreceptor cell death in degenerating retinas, research has also focused on elevating the expression of the most extensively characterized natural apoptotic inhibitor, XIAP, or X-linked inhibitor of apoptosis. This protein, as its name implies, is encoded by a gene on chromsome X, binding through its baculovirus-inhibitor of apoptosis repeat (BIR) domains, caspases 3, 7 and 9, thus inhibiting the process of apoptosis [81]. Many studies have been reported in which the protective effects of XIAP expression have been assessed in animal models of retinal degeneration, glaucoma and retinal detachment and on RPE cells in relation to oxidative stress [82–86]. In addition, XIAP expression has recently been shown to enhance the viability of transplanted photoreceptor precursor cells in mice [87]. In a study reported by Leonard and colleagues [85], the potentially protective effects of XIAP expression were assessed in two well characterized rat models of autosomal dominant RP, expressing either P23H or S334ter mutations of the rhodopsin gene. In this series of experiments, a full length XIAP gene was incorporated into the genome of an AAV5 vector under the control of a chicken ß-actin promoter. The virus was introduced into the retinas of these animals by sub-retinal inoculation. Animals were injected between post-natal days 14 and 17 and were functionally and structurally assessed 28 weeks after inoculation. The ONL of treated animals was significantly thicker than controls in both animal models. Functional assessment of the retinas of these animals was also undertaken by electroretinography. Little or no improvement in ERG waveforms was observed in S334ter animals, although both ERG a- and b-wave amplitudes were significantly improved in animals with the P23H transgene. The authors conclude that their results 'clearly demonstrate the potential of XIAP gene therapy for the treatment of inherited retinal degenerations. It is important to note [however] that the XIAP treatment does not represent an all-encompassing gene therapy strategy for retinal degeneration'. In our opinion, XIAP therapy, based on these data, could be potentially effectively used in combination with therapy targeting specifically-mutated genes.

Interestingly, XIAP therapy has also recently been used by Yao and colleagues to enhance the viability of transplanted photoreceptor precursors in a murine model [87]. MacLaren and colleagues initially demonstrated in 2006, that transplantation of photoreceptor precursors from donor mice at the time of peak of rod photoreceptor genesis (post natal day 1) resulted in the development of transplanted precursors into apparently mature photoreceptors that were able to migrate into the outer nuclear layer of transplanted retinas [88]. Yao and colleagues pre-transfected immature photoreceptor cells with a XIAP-expressing AAV vector and were able to show that transplanted cells showed increased survival over those that were not expressing XIAP, the authors concluding that 'preventing programmed cell death through XIAP therapy may be an important component of future therapeutic retinal cell transplantation strategies'. It is of interest also to note that expression of XIAP has also been shown by Shan and colleagues to protect retinal RPE cells from apoptosis induced by exposure of such cells to peroxide [89]. The authors suggest that expression of XIAP may, therefore, be beneficial in protecting the RPE and photoreceptors from oxidative stress in age-related macular degeneration.

In terms addressing the concept of improving the efficiency of the apoptosis in degenerating retinas to reduce the likelihood of exacerbating retinal degeneration owing to cell disruption by necrotic processes, we have recently shown that cone photoreceptor survival in mice homozygous for a targeted mutation of the rhodopsin gene, a model of autosomal recessive RP, is significantly enhanced by the presence of C1q, a primary component of the classical complement pathway [90]. The rationale for addressing this hypothesis was that C1q is known to bind to cells in late-stages of apoptosis, which results in deposition onto such cells of complement components C3 and C4, facilitating their phagocytosis. In the absence of C1q, the efficiency of apoptotic clearance of dying photoreceptors may be reduced, favoring necrosis and cell lysis and the spilling of cell contents into the retina, which may negatively impact on photoreceptor survival. These results suggest that optimum cone cell survival and hence daytime vision, will be achieved by maintaining, or possibly enhancing levels of C1q within the retina. It is also of interest to note that Galvan and colleagues have recently shown that C1q enhances the expression within the retina of the MERTK which is active in facilitating engulfment of shed outer segment photoreceptor disk membranes into the RPE, providing an additional possible explanation for the protective effects of C1q on photoreceptor survival [91].

In conclusion, gene therapies are now in human clinical trial for some forms of degenerative retinopathy, while an impressive portfolio of gene-based medicines targeting primary genetic lesions, in addition to those targeting disease pathologies common to multiple forms of disease have now been convincingly validated in a growing number of animal disease models, setting the scene for translation into the clinical setting.

References

1. Acland GM, Aguirre GD, Ray J, Zhang Q, Aleman TS, Cideciyan AV, Pearce-Kelling SE, Anand V, Zeng Y, Maguire AM, Jacobson SG, Hauswirth WW, Bennett J (2001) Gene therapy restores vision in a canine model of childhood blindness. Nat Genet 28(1):92–95
2. D'Cruz PM, Yasumura D, Weir J, Matthes MT, Abderrahim H, LaVail MM, Vollrath D (2000) Mutation of the receptor tyrosine kinase gene Mertk in the retinal dystrophic RCS rat. Hum Mol Genet 9(4):645–651
3. Gal A, Li Y, Thompson DA, Weir J, Orth U, Jacobson SG, Apfelstedt-Sylla E, Vollrath D (2000) Mutations in MERTK, the human orthologue of the RCS rat retinal dystrophy gene, cause retinitis pigmentosa. Nat Genet 26(3):270–271
4. Smith AJ, Schlichtenbrede FC, Tschernutter M, Bainbridge JW, Thrasher AJ, Ali RR (2003) AAV-mediated gene transfer slows photoreceptor loss in the RCS rat model of retinitis pigmentosa. Mol Ther 8(2):188–195

5. Tschernutter M, Schlichtenbrede FC, Howe S, Balaggan KS, Munro PM, Bainbridge JW, Thrasher AJ, Smith AJ, Ali RR (2005) Long-term preservation of retinal function in the RCS rat model of retinitis pigmentosa following lentivirus-mediated gene therapy. Gene Ther 12(8):694–701

6. Gibson F, Walsh J, Mburu P, Varela A, Brown KA, Antonio M, Beisel KW, Steel KP, Brown SD (1995) A type VII myosin encoded by the mouse deafness gene shaker-1. Nature 374(6517):62–64

7. Weil D, Blanchard S, Kaplan J, Guilford P, Gibson F, Walsh J, Mburu P, Varela A, Levilliers J, Weston MD, Kelley PM, Kimberling WJ, Wagenaar M, Levi-Acobas F, Larget-Piet D, Munnich A, Steel KP, Brown SDM, Petit C (1995) Defective myosin VIIA gene responsible for Usher syndrome type 1B. Nature 374(6517):60–61

8. Hashimoto T, Gibbs D, Lillo C, Azarian SM, Legacki E, Zhang XM, Yang XJ, Williams DS (2007) Lentiviral gene replacement therapy of retinas in a mouse model for Usher syndrome type 1B. Gene Ther 14(7):584–594

9. Moiseyev G, Nikolaeva O, Chen Y, Farjo K, Takahashi Y, Ma JX (2010) Inhibition of the visual cycle by A2E through direct interaction with RPE65 and implications in Stargardt disease. Proc Natl Acad Sci U S A 107(41):17551–17556

10. Allikmets R, Shroyer NF, Singh N, Seddon JM, Lewis RA, Bernstein PS, Peiffer A, Zabriskie NA, Li Y, Hutchinson A, Dean M, Lupski JR, Leppert M (1997) Mutation of the Stargardt disease gene (ABCR) in age-related macular degeneration. Science 277(5333):1805–1807

11. Kong J, Kim SR, Binley K, Pata I, Doi K, Mannik J, Zernant-Rajang J, Kan O, Iqball S, Naylor S, Sparrow JR, Gouras P, Allikmets R (2008) Correction of the disease phenotype in the mouse model of Stargardt disease by lentiviral gene therapy. Gene Ther 15(19):1311–1320

12. Pawlyk BS, Smith AJ, Buch PK, Adamian M, Hong DH, Sandberg MA, Ali RR, Li T (2005) Gene replacement therapy rescues photoreceptor degeneration in a murine model of Leber congenital amaurosis lacking RPGRIP. Invest Ophthalmol Vis Sci 46(9):3039–3045

13. Min SH, Molday LL, Seeliger MW, Dinculescu A, Timmers AM, Janssen A, Tonagel F, Tanimoto N, Weber BH, Molday RS, Hauswirth WW (2005) Prolonged recovery of retinal structure/function after gene therapy in an Rs1h-deficient mouse model of x-linked juvenile retinoschisis. Mol Ther 12(4):644–651

14. Kjellstrom S, Bush RA, Zeng Y, Takada Y, Sieving PA (2007) Retinoschisin gene therapy and natural history in the Rs1h-KO mouse: long-term rescue from retinal degeneration. Invest Ophthalmol Vis Sci 48(8):3837–3845

15. Park TK, Wu Z, Kjellstrom S, Zeng Y, Bush RA, Sieving PA, Colosi P (2009) Intravitreal delivery of AAV8 retinoschisin results in cell type-specific gene expression and retinal rescue in the Rs1-KO mouse. Gene Ther 16(7):916–926. Erratum in: Gene Ther 16(7):941 (2009)

16. Alexander JJ, Umino Y, Everhart D, Chang B, Min SH, Li Q, Timmers AM, Hawes NL, Pang JJ, Barlow RB, Hauswirth WW (2007) Restoration of cone vision in a mouse model of achromatopsia. Nat Med 13(6):685–687

17. Komáromy AM, Alexander JJ, Rowlan JS, Garcia MM, Chiodo VA, Kaya A, Tanaka JC, Acland GM, Hauswirth WW, Aguirre GD (2010) Gene therapy rescues cone function in congenital achromatopsia. Hum Mol Genet 19(13):2581–2593. Erratum in: Hum Mol Genet 20(24):5024 (2011)

18. Carvalho LS, Xu J, Pearson RA, Smith AJ, Bainbridge JW, Morris LM, Fliesler SJ, Ding XQ, Ali RR (2011) Long-term and age-dependent restoration of visual function in a mouse model of CNGB3-associated achromatopsia following gene therapy. Hum Mol Genet 20(16):3161–3175

19. Michalakis S, Mühlfriedel R, Tanimoto N, Krishnamoorthy V, Koch S, Fischer MD, Becirovic E, Bai L, Huber G, Beck SC, Fahl E, Büning H, Paquet-Durand F, Zong X, Gollisch T, Biel M, Seeliger MW (2010) Restoration of cone vision in the CNGA3−/− mouse model of congenital complete lack of cone photoreceptor function. Mol Ther 18(12):2057–2063

20. Pang JJ, Boye SL, Kumar A, Dinculescu A, Deng W, Li J, Li Q, Rani A, Foster TC, Chang B, Hawes NL, Boatright JH, Hauswirth WW (2008) AAV-mediated gene therapy for retinal degeneration in the rd10 mouse containing a recessive PDEbeta mutation. Invest Ophthalmol Vis Sci 49(10):4278–4283

21. Pang JJ, Dai X, Boye SE, Barone I, Boye SL, Mao S, Everhart D, Dinculescu A, Liu L, Umino Y, Lei B, Chang B, Barlow R, Strettoi E, Hauswirth WW (2011) Long-term retinal function and structure rescue using capsid mutant AAV8 vector in the rd10 mouse, a model of recessive retinitis pigmentosa. Mol Ther 19(2):234–242

22. Tan MH, Smith AJ, Pawlyk B, Xu X, Liu X, Bainbridge JB, Basche M, McIntosh J, Tran HV, Nathwani A, Li T, Ali RR (2009) Gene therapy for retinitis pigmentosa and Leber congenital amaurosis caused by defects in AIPL1: effective rescue of mouse models of partial and complete Aipl1 deficiency using AAV2/2 and AAV2/8 vectors. Hum Mol Genet 18(12):2099–2114. Erratum in: Hum Mol Genet 19(4):735 (2010)

23. Sun X, Pawlyk B, Xu X, Liu X, Bulgakov OV, Adamian M, Sandberg MA, Khani SC, Tan MH, Smith AJ, Ali RR, Li T (2010) Gene therapy with a promoter targeting both rods and cones rescues retinal degeneration caused by AIPL1 mutations. Gene Ther 17(1):117–131

24. Pawlyk BS, Bulgakov OV, Liu X, Xu X, Adamian M, Sun X, Khani SC, Berson EL, Sandberg MA, Li T (2010) Replacement gene therapy with a human RPGRIP1 sequence slows photoreceptor degeneration in a murine model of Leber congenital amaurosis. Hum Gene Ther 21(8):993–1004

25. Boye SE, Boye SL, Pang J, Ryals R, Everhart D, Umino Y, Neeley AW, Besharse J, Barlow R, Hauswirth WW (2010) Functional and behavioral restoration of vision by gene therapy in the guanylate cyclase-1 (GC1) knockout mouse. PLoS ONE 5(6):e11306

26. Mihelec M, Pearson RA, Robbie SJ, Buch PK, Azam SA, Bainbridge JW, Smith AJ, Ali RR (2011) Long-term preservation of cones and improvement in visual function following gene therapy in a mouse model of Leber congenital amaurosis caused by guanylate cyclase-1 deficiency. Hum Gene Ther 22(10):1179–1190

27. Zou J, Luo L, Shen Z, Chiodo VA, Ambati BK, Hauswirth WW, Yang J (2011) Whirlin replacement restores the formation of the USH2 protein complex in whirlin knockout photoreceptors. Invest Ophthalmol Vis Sci 52(5):2343–2351

28. Dinculescu A, Estreicher J, Zenteno JC, Aleman TS, Schwartz SB, Huang WC, Roman AJ, Sumaroka A, Li Q, Deng WT, Min SH, Chiodo VA, Neeley A, Liu X, Shu X, Matias-Florentino M, Buentello-Volante B, Boye SL, Cideciyan AV, Hauswirth WW, Jacobson SG (2012) Gene therapy for Retinitis Pigmentosa caused by MFRP mutations: human phenotype and preliminary proof of concept. Hum Gene Ther 23(4):367–376

29. Simons DL, Boye SL, Hauswirth WW, Wu SM (2011) Gene therapy prevents photoreceptor death and preserves retinal function in a Bardet-Biedl syndrome mouse model. Proc Natl Acad Sci U S A 108(15):6276–6281

30. Millington-Ward S, O'Neill B, Tuohy G, Al-Jandal N, Kiang AS, Kenna PF, Palfi A, Hayden P, Mansergh F, Kennan A, Humphries P, Farrar GJ (1997) Strategems in vitro for gene therapies directed to dominant mutations. Hum Mol Genet 6(9):1415–1426

31. Farrar GJ, Kenna PF, Humphries P (2002) On the genetics of retinitis pigmentosa and on mutation-independent approaches to therapeutic intervention. EMBO J 21(5):857–864

32. O'Reilly M, Palfi A, Chadderton N, Millington-Ward S, Ader M, Cronin T, Tuohy T, Auricchio A, Hildinger M, Tivnan A, McNally N, Humphries MM, Kiang AS, Humphries P, Kenna PF, Farrar GJ (2007) RNA interference-mediated suppression and replacement of human rhodopsin in vivo. Am J Hum Genet 81(1):127–135

33. Chadderton N, Millington-Ward S, Palfi A, O'Reilly M, Tuohy G, Humphries MM, Li T, Humphries P, Kenna PF, Farrar GJ (2009) Improved retinal function in a mouse model of dominant retinitis pigmentosa following AAV-delivered gene therapy. Mol Ther 17(4):593–599

34. Palfi A, Millington-Ward S, Chadderton N, O'Reilly M, Goldmann T, Humphries MM, Li T, Wolfrum U, Humphries P, Kenna PF, Farrar GJ (2010) Adeno-associated virus-mediated rhodopsin replacement provides therapeutic benefit in mice with a targeted disruption of the rhodopsin gene. Hum Gene Ther 21(3):311–323

35. Millington-Ward S, Chadderton N, O'Reilly M, Palfi A, Goldmann T, Kilty C, Humphries M, Wolfrum U, Bennett J, Humphries P, Kenna PF, Farrar GJ (2011) Suppression and replacement gene therapy for autosomal dominant disease in a murine model of dominant retinitis pigmentosa. Mol Ther 19(4):642–649

36. Mao H, James T Jr, Schwein A, Shabashvili AE, Hauswirth WW, Gorbatyuk MS, Lewin AS (2011) AAV delivery of wild-type rhodopsin preserves retinal function in a mouse model of autosomal dominant retinitis pigmentosa. Hum Gene Ther 22(5):567–575

37. Aherne A, Kennan A, Kenna PF, McNally N, Lloyd DG, Alberts IL, Kiang AS, Humphries MM, Ayuso C, Engel PC, Gu JJ, Mitchell BS, Farrar GJ, Humphries P (2004) On the molecular pathology of neurodegeneration in IMPDH1-based retinitis pigmentosa. Hum Mol Genet 13(6):641–650

38. Tam LC, Kiang AS, Kennan A, Kenna PF, Chadderton N, Ader M, Palfi A, Aherne A, Ayuso C, Campbell M, Reynolds A, McKee A, Humphries MM, Farrar GJ, Humphries P (2008) Therapeutic benefit derived from RNAi-mediated ablation of IMPDH1 transcripts in a murine model of autosomal dominant retinitis pigmentosa (RP10). Hum Mol Genet 17(14):2084–2100

39. Bi A, Cui J, Ma YP, Olshevskaya E, Pu M, Dizhoor AM, Pan ZH (2006) Ectopic expression of a microbial-type rhodopsin restores visual responses in mice with photoreceptor degeneration. Neuron 50(1):23–33

40. Tomita H, Sugano E, Fukazawa Y, Isago H, Sugiyama Y, Hiroi T, Ishizuka T, Mushiake H, Kato M, Hirabayashi M, Shigemoto R, Yawo H, Tamai M (2009) Visual properties of transgenic rats harboring the channelrhodopsin-2 gene regulated by the thy-1.2 promoter. PLoS ONE 4(11):e7679

41. Tomita H, Sugano E, Isago H, Hiroi T, Wang Z, Ohta E, Tamai M (2010) Channelrhodopsin-2 gene transduced into retinal ganglion cells restores functional vision in genetically blind rats. Exp Eye Res 90(3):429–436. Epub 2009 Dec 27

42. Busskamp V, Duebel J, Balya D, Fradot M, Viney TJ, Siegert S, Groner AC, Cabuy E, Forster V, Seeliger M, Biel M, Humphries P, Paques M, Mohand-Said S, Trono D, Deisseroth K, Sahel JA, Picaud S, Roska B (2010) Genetic reactivation of cone photoreceptors restores visual responses in retinitis pigmentosa. Science 329(5990):413–417

43. Faktorovich EG, Steinberg RH, Yasumura D, Matthes MT, LaVail MM (1990) Photoreceptor degeneration in inherited retinal dystrophy delayed by basic fibroblast growth factor. Nature 347(6288):83–86

44. LaVail MM, Unoki K, Yasumura D, Matthes MT, Yancopoulos GD, Steinberg RH (1992) Multiple growth factors, cytokines, and neurotrophins rescue photoreceptors from the damaging effects of constant light. Proc Natl Acad Sci U S A 89(23):11249–11253

45. LaVail MM, Yasumura D, Matthes MT, Lau-Villacorta C, Unoki K, Sung CH, Steinberg RH (1998) Protection of mouse photoreceptors by survival factors in retinal degenerations. Invest Ophthalmol Vis Sci 39(3):592–602

46. Whiteley SJ, Klassen H, Coffer PJ, Young MJ (2001) Photoreceptor rescue after low-dose intravitreal IL-1ß injection in the RCS rat. Exp Eye Res 73(4):557–568

47. Cayouette M, Gravel C (1997) Adenovirus-mediated gene transfer of ciliary neurotrophic factor can prevent photoreceptor degeneration in the retinal degeneration (rd) mouse. Hum Gene Ther 8(4):423–430

48. Liang FQ, Dejneka NS, Cohen DR, Krasnoperova NV, Lem J, Maguire AM, Dudus L, Fisher KJ, Bennett J (2001) AAV-mediated delivery of ciliary neurotrophic factor prolongs photoreceptor survival in the rhodopsin knockout mouse. Mol Ther 3(2):241–248

49. Liang FQ, Aleman TS, Dejneka NS, Dudus L, Fisher KJ, Maguire AM, Jacobson SG, Bennett J (2001) Long-term protection of retinal structure but not function using RAAV.CNTF in animal models of retinitis pigmentosa. Mol Ther 4(5):461–472

50. Bok D, Yasumura D, Matthes MT, Ruiz A, Duncan JL, Chappelow AV, Zolutukhin S, Hauswirth W, LaVail MM (2002) Effects of adeno-associated virus-vectored ciliary neurotrophic factor on retinal structure and function in mice with a P216L rds/peripherin mutation. Exp Eye Res 74(6):719–735

51. Schlichtenbrede FC, MacNeil A, Bainbridge JW, Tschernutter M, Thrasher AJ, Smith AJ, Ali RR (2003) Intraocular gene delivery of ciliary neurotrophic factor results in significant loss of retinal function in normal mice and in the Prph2Rd2/Rd2 model of retinal degeneration. Gene Ther 10(6):523–527
52. Adamus G, Sugden B, Shiraga S, Timmers AM, Hauswirth WW (2003) Anti-apoptotic effects of CNTF gene transfer on photoreceptor degeneration in experimental antibody-induced retinopathy. J Autoimmun 21(2):121–129. Erratum in: J Autoimmun 21(4):393 (2003)
53. Huang SP, Lin PK, Liu JH, Khor CN, Lee YJ (2004) Intraocular gene transfer of ciliary neurotrophic factor rescues photoreceptor degeneration in RCS rats. J Biomed Sci 11(1):37–48
54. Buch PK, MacLaren RE, Durán Y, Balaggan KS, MacNeil A, Schlichtenbrede FC, Smith AJ, Ali RR (2006) In contrast to AAV-mediated Cntf expression, AAV-mediated Gdnf expression enhances gene replacement therapy in rodent models of retinal degeneration. Mol Ther 14(5):700–709
55. Talcott KE, Ratnam K, Sundquist SM, Lucero AS, Lujan BJ, Tao W, Porco TC, Roorda A, Duncan JL (2011) Longitudinal study of cone photoreceptors during retinal degeneration and in response to ciliary neurotrophic factor treatment. Invest Ophthalmol Vis Sci 52(5):2219–2226
56. Frasson M, Picaud S, Léveillard T, Simonutti M, Mohand-Said S, Dreyfus H, Hicks D, Sabel J (1999) Glial cell line-derived neurotrophic factor induces histologic and functional protection of rod photoreceptors in the rd/rd mouse. Invest Ophthalmol Vis Sci 40(11):2724–2734
57. McGee Sanftner LH, Abel H, Hauswirth WW, Flannery JG (2001) Glial cell line derived neurotrophic factor delays photoreceptor degeneration in a transgenic rat model of retinitis pigmentosa. Mol Ther 4(6):622–629
58. Wu WC, Lai CC, Chen SL, Sun MH, Xiao X, Chen TL, Lin KK, Kuo SW, Tsao YP (2005) Long-term safety of GDNF gene delivery in the retina. Curr Eye Res 30(8):715–722
59. Hauck SM, Kinkl N, Deeg CA, Swiatek-de Lange M, Schöffmann S, Ueffing M (2006) GDNF family ligands trigger indirect neuroprotective signaling in retinal glial cells. Mol Cell Biol 26(7):2746–2757
60. Dong A, Shen J, Krause M, Hackett SF, Campochiaro PA (2007) Increased expression of glial cell line-derived neurotrophic factor protects against oxidative damage-induced retinal degeneration. J Neurochem 103(3):1041–1052
61. Gregory-Evans K, Chang F, Hodges MD, Gregory-Evans CY (2009) Ex vivo gene therapy using intravitreal injection of GDNF-secreting mouse embryonic stem cells in a rat model of retinal degeneration. Mol Vis 15:962–973
62. Del Río P, Irmler M, Arango-González B, Favor J, Bobe C, Bartsch U, Vecino E, Beckers J, Hauck SM, Ueffing M (2011) GDNF-induced osteopontin from Müller glial cells promotes photoreceptor survival in the Pde6brd1 mouse model of retinal degeneration. Glia 59(5): 821–832
63. Dalkara D, Kolstad KD, Guerin KI, Hoffmann NV, Visel M, Klimczak RR, Schaffer DV, Flannery JG (2011) AAV mediated GDNF secretion from retinal glia slows down retinal degeneration in a rat model of retinitis pigmentosa. Mol Ther 19(9):1602–1608
64. Klimczak RR, Koerber JT, Dalkara D, Flannery JG, Schaffer DV (2009) A novel adeno-associated viral variant for efficient and selective intravitreal transduction of rat Müller cells. PLoS ONE 4(10):e7467
65. Mohand-Said S, Hicks D, Simonutti M, Tran-Minh D, Deudon-Combe A, Dreyfus H, Silverman MS, Ogilvie JM, Tenkova T, Sahel J (1997) Photoreceptor transplants increase host cone survival in the retinal degeneration (rd) mouse. Ophthalmic Res 29(5):290–297
66. Mohand-Said S, Deudon-Combe A, Hicks D, Simonutti M, Forster V, Fintz AC, Léveillard T, Dreyfus H, Sahel JA (1998) Normal retina releases a diffusible factor stimulating cone survival in the retinal degeneration mouse. Proc Natl Acad Sci U S A 95(14):8357–8362
67. Mohand-Said S, Hicks D, Dreyfus H, Sahel JA (2000) Selective transplantation of rods delays cone loss in a retinitis pigmentosa model. Arch Ophthalmol 118(6):807–811

68. Fintz AC, Audo I, Hicks D, Mohand-Said S, Léveillard T, Sahel J (2003) Partial characterization of retina-derived cone neuroprotection in two culture models of photoreceptor degeneration. Invest Ophthalmol Vis Sci 44(2):818–825
69. Léveillard T, Mohand-Sa S, Lorentz O, Hicks D, Fintz AC, Clérin E, Simonutti M, Forster V, Cavusoglu N, Chalmel F, Dollé P, Poch O, Lambrou G, Sahel JA (2004) Identification and characterization of rod-derived cone viability factor. Nat Genet 36(7):755–759
70. Léveillard T, Sahel JA (2010) Rod-derived cone viability factor for treating blinding diseases: from the clinic to redox signaling. Sci Transl Med 2(26):26ps16
71. Yang Y, Mohand-Said S, Danan A, Simonutti M, Fontaine V, Clerin E, Picaud S, Léveillard T, Sahel JA (2009) Functional cone rescue by RdCVF protein in a dominant model of retinitis pigmentosa. Mol Ther 17(5):787–795
72. Chang GQ, Hao Y, Wong F (1993) Apoptosis: final common pathway of photoreceptor death in rd, rds, and rhodopsin mutant mice. Neuron 11(4):595–605
73. Lolley RN, Rong H, Craft CM (1994) Linkage of photoreceptor degeneration by apoptosis with inherited defect in phototransduction. Invest Ophthalmol Vis Sci 35(2):358–362
74. Portera-Cailliau C, Sung CH, Nathans J, Adler R (1994) Apoptotic photoreceptor cell death in mouse models of retinitis pigmentosa. Proc Natl Acad Sci U S A 91(3):974–978
75. Joseph RM, Li T (1996) Overexpression of Bcl-2 or Bcl-XL transgenes and photoreceptor degeneration. Invest Ophthalmol Vis Sci 37(12):2434–2446
76. Chen J, Flannery JG, LaVail MM, Steinberg RH, Xu J, Simon MI (1996) bcl-2 overexpression reduces apoptotic photoreceptor cell death in three different retinal degenerations. Proc Natl Acad Sci U S A 93(14):7042–7047
77. Tsang SH, Chen J, Kjeldbye H, Li WS, Simon MI, Gouras P, Goff SP (1997) Retarding photoreceptor degeneration in Pdegtm1/Pdegtml mice by an apoptosis suppressor gene. Invest Ophthalmol Vis Sci 38(5):943–950
78. Nir I, Kedzierski W, Chen J, Travis GH (2000) Expression of Bcl-2 protects against photoreceptor degeneration in retinal degeneration slow (rds) mice. J Neurosci 20(6): 2150–2154
79. Kluck RM, Bossy-Wetzel E, Green DR, Newmeyer DD (1997) The release of cytochrome c from mitochondria: a primary site for Bcl-2 regulation of apoptosis. Science 275(5303):1132–1136
80. Yang J, Liu X, Bhalla K, Kim CN, Ibrado AM, Cai J, Peng TI, Jones DP, Wang X (1997) Prevention of apoptosis by Bcl-2: release of cytochrome c from mitochondria blocked. Science 275(5303):1129–1132
81. Holcik M, Gibson H, Korneluk RG (2001) XIAP: apoptotic brake and promising therapeutic target. Apoptosis 6(4):253–261
82. Kügler S, Straten G, Kreppel F, Isenmann S, Liston P, Bähr M (2000) The X-linked inhibitor of apoptosis (XIAP) prevents cell death in axotomized CNS neurons in vivo. Cell Death Differ 7(9):815–824
83. McKinnon SJ, Lehman DM, Tahzib NG, Ransom NL, Reitsamer HA, Liston P, LaCasse E, Li Q, Korneluk RG, Hauswirth WW (2002) Baculoviral IAP repeat-containing-4 protects optic nerve axons in a rat glaucoma model. Mol Ther 5(6):780–787
84. Zadro-Lamoureux LA, Zacks DN, Baker AN, Zheng QD, Hauswirth WW, Tsilfidis C (2009) XIAP effects on retinal detachment-induced photoreceptor apoptosis [corrected]. Invest Ophthalmol Vis Sci 50(3):1448–1453. Erratum in Invest Ophthalmol Vis Sci 50(4):1530 (2009)
85. Leonard KC, Petrin D, Coupland SG, Baker AN, Leonard BC, LaCasse EC, Hauswirth WW, Korneluk RG, Tsilfidis C (2007) XIAP protection of photoreceptors in animal models of retinitis pigmentosa. PLoS ONE 2(3):e314
86. Petrin D, Baker A, Coupland SG, Liston P, Narang M, Damji K, Leonard B, Chiodo VA, Timmers A, Hauswirth W, Korneluk RG, Tsilfidis C (2003) Structural and functional protection of photoreceptors from MNU-induced retinal degeneration by the X-linked inhibitor of apoptosis. Invest Ophthalmol Vis Sci 44(6):2757–2763

87. Yao J, Feathers KL, Khanna H, Thompson D, Tsilfidis C, Hauswirth WW, Heckenlively JR, Swaroop A, Zacks DN (2011) XIAP therapy increases survival of transplanted rod precursors in a degenerating host retina. Invest Ophthalmol Vis Sci 52(3):1567–1572

88. MacLaren RE, Pearson RA, MacNeil A, Douglas RH, Salt TE, Akimoto M, Swaroop A, Sowden JC, Ali RR (2006) Retinal repair by transplantation of photoreceptor precursors. Nature 444(7116):203–207

89. Shan H, Ji D, Barnard AR, Lipinski DM, You Q, Lee EJ, Kamalden TA, Sun X, MacLaren RE (2011) AAV-mediated gene transfer of human X-linked inhibitor of apoptosis protects against oxidative cell death in human RPE cells. Invest Ophthalmol Vis Sci 52(13):9591–9597

90. Humphries MM, Kenna PF, Campbell M, Tam LC, Nguyen AT, Farrar GJ, Botto M, Kiang AS, Humphries P (2012) C1q enhances cone photoreceptor survival in a mouse model of autosomal recessive retinitis pigmentosa. Eur J Hum Genet 20(1):64–68

91. Galvan MD, Foreman DB, Zeng E, Tan JC, Bohlson SS (2012) Complement component c1q regulates macrophage expression of mer tyrosine kinase to promote clearance of apoptotic cells. J Immunol 188(8):3716–3723

Chapter 3
Molecular Medicines

Clinical trials that have either been registered or are currently recruiting for RP and LCA include assessment of efficacies of a number of orally administered compounds including valproic acid, docosahexeanoic acid (DHA), vitamins A and E, lutein and a 9-cis retinal enriched compound. These include a trial of oral valproic acid for retinitis pigmentosa, phase II (NCT01233609, National Neurovision Research Institute), a randomized clinical trail for retinitis pigmentosa, phase II, involving dietary supplementation with vitamin A palmitate (NCT00346333, National Eye Institute), treatment with DHA (docosahexanoic acid) in X-linked retinitis pigmentosa, phase II (NCT00100230, Foundation Fighting Blindness and Martek Biosciences Corporation), on the efficacy and safety of oral valproic acid for retinitis pigmentosa, phase II (NCT01399515, Seoul National University Hospital), a study of the effects of oral administration of 9-cis rich powder of the alga Dunaliella Bardawil on visual functions in patients with autosomal dominant retinitis pigmentosa, (NCT01256697, Sheba Medical Centre), and a safety/proof of concept study of oral QLT1001 (a medication based on systemic delivery of 9-cis retinal) in subjects with LCA or RP due to RPE65 or LRAT mutations (NCT01014052, QLT inc., Vancouver, November 12 2009).

In regard to valproic acid as a potential therapeutic for RP, the results of a limited trial of this drug were published in 2010 by Clemson et al. [1]. In that study, seven patients with RP with undefined causative mutations were administered between 500 and 750 mg/day of the agent for a period of 4 months, readouts of therapeutic efficacy being visual field and visual acuity. This drug is an FDA-approved anticonvulsant and is also used in mood stabilization. One of the reasons the authors considered this drug for treatment of RP was because of their observations that it increases the yield of properly folded mutated rhodopsin proteins. In total 13 eyes were examined, nine showing improvement, two showing no improvement and two showing a decrease in visual fields. The authors cautiously conclude: 'Treatment with VPA is suggestive of a therapeutic benefit to patients with RP. A placebo-controlled clinical trial will be necessary to assess the efficacy and safety of VPA for RP rigorously'. As noted above, two clinical trials are now in progress.

P. Humphries et al., *Hereditary Retinopathies*, SpringerBriefs in Genetics, DOI: 10.1007/978-1-4614-4499-2_3, © The Author(s) 2012

Many studies have been reported in which the effects of vitamin supplementation have been assessed in regard to slowing the progression of RP. Retinal, the oxidized form of vitamin A is essential in vision. Photoreceptor opsins bind the chromophore, 11-cis retinal. Following light absorption, retinal isomerizes to the all-trans form and dissociates from opsin, entering the underlying pigmented epithelium, where it is recycled by the retinoid (visual) cycle back into the 11-cis form before being transported back into the photoreceptor cells and binding again as the chromophore, to opsin molecules. Dietary supplementation of vitamin A precursors to fighter pilots (feeding them carrots rich in the vitamin A precursor, ß-carotene) as an aid to night vision was used as a subterfuge by the British Government during world war II to conceal the increasingly effective power of radar in destroying enemy aircraft, an 'urban myth' that has persisted to this day. Major studies of dietary supplementation of vitamins in slowing the progression of retinitis pigmentosa have been reported [2–7]. In the study by Berson et al. [2], the effects of high dosages of vitamin A and vitamin E (15,000 and 400 IU/day respectively) were evaluated, the authors concluding that consumption of vitamin A at this daily level may be beneficial in slowing down the progression of RP. In a second study by Berson et al. [5] dietary supplementation with the omega fatty acid docosahexaenoic acid (DHA) was suggested to improve the benefit of high dose vitamin A supplementation. In a more recent study by Berson et al. [7], patients receiving high dosage vitamin A supplement were also given 12 mg of Lutein per day. Lutein, together with zeaxanthin are two macular pigments which may protect the retina against oxygen stress. They cannot be synthesized by the body and must be obtained from green or yellow vegetables. The conclusion from this trial was that 'lutein supplementation of 12 mg/ml slowed loss of midperipheral visual field on average among nonsmoking adults with retinitis pigmentosa taking vitamin A'. In an accompanying Editorial [8], Drs. Massof and Fishman conclude, 'The three trials reviewed here [and referred to above] are examples of well-executed studies and they present a number of provocative results. However, none of these studies convincingly proved that the treatments put on trial are effective in slowing the rate of progression of RP and, therefore, do not warrant mandating a change in how patients with RP are treated'.

While up to 15 genes have been associated with Leber congenital amaurosis (see Table 1, Chap. 1), up to 10 % of cases of this disease are caused by mutations in the RPE65 gene, encoding an enzymatic component of the visual cycle, defects in which cause a block in the regeneration of 11-cis retinal and an accumulation of all-trans retinyl esters in the RPE. Van Hooser et al. [9] investigated the concept of oral administration of the analoge, 9-cis retinal to mice with a targeted disruption of the RPE65 gene. 9-cis retinal was used because it is more stable than the 11-cis form, while still being capable of forming isorhodopsin and when administered by oral gavage to 8–12 week old *RPE65-/-* mice, improvements in electroretinographic function were observed within 48 h of treatment. The authors indicate, 'Within 48 h, there was formation of rod photopigment and dramatic improvement in rod physiology, thus demonstrating that mechanism-based pharmacological intervention has the potential to restore vision in otherwise incurable genetic retinal degenerations'. In a subsequent study, Van Hooser et al. were able to demonstrate that early treatment in these animals resulted in long-term improvement in visual function, for at least 6 months and concluded that such treatment may form a useful therapy for LCA caused by null mutations within the RPE65 gene [10]. As outlined above, clinical trials showing positive preliminary results are now in progress involving oral administration of QLT091001, an orally available 11-cis retinal replacement (http://www.qltinc.com).

Retinals have also proven useful as pharmacological chaperones by stabilizing mutant opsins and assisting in improved folding and trafficking of mutant P23H protein expressed in cell cultures [11], although it was unclear as to whether this strategy was worth pursuing in the P23H diseased retina where 11-cis retinal is found in abundance. However, a recent report from Palczewski et al. demonstrated in P23H knockin mice (*P23H/+*) that genetic ablation of 11-cis retinal by crossing the knockin mice onto an *lrat-/-* background (*P23H/+ lrat-/-*) hastened structural degeneration at month 1 when compared to *P23H/+* mice even though no such degeneration was observed in either *rho +/+lrat-/-* or *rho+/-/lrat-/-* animals at that timepoint [12]. This observation suggests that a lack of 11-cis retinal can increase the toxicity of the P23H mutation during development and this taken together with the above described studies on improved folding and trafficking of P23H by retinals, suggests that therapeutic application of 9-cis retinal as in the LCA trial, could conceivably retard retinal degeneration in patients carrying the P23H opsin mutation. Intriguingly, the same therapeutic may even work for other RP patients carrying mutations in genes expressing proteins which either mislocalize or are involved in retinal cycling, including the recently identified dominant RPE65 mutation D477G, which causes RP with choroidal involvement [13].

In regard to small molecule inducers of ER and cytosolic stress responses inhibition of hsp90 by the ansamycin antibiotic geldanamycin or its less toxic derivatives 17-AAG (tanespimycin) and 17-DMAG (alvespimycin) has two major effects on the cell. Firstly, by binding to the ATP pocket of hsp90 the release of HSF-1 from cytoplasm to nucleus is stimulated thus upregulating the transcription of heat shock response proteins and secondly, the binding and thus maturation of hsp90 client proteins to hsp90 is prevented. The latter effect makes hsp90 inhibition an attractive target in multiple cancers since many proto-oncogenic proteins which become deregulated in tumors and VEGF receptors which are involved in angiogenesis are client proteins of hsp90 [14–17]. In fact, 17-AAG currently has 49 cancer clinical trials associated with it while in a phase II study on advanced solid tumors, patients are being treated intravenously (80 mg/m^2/week) with the water soluble, orally available derivative, 17-DMAG. However, it is the effect on protein misfolding via induction of the unfolded protein response in the ER [18] and cytoplasmic heat shock response [19] of this class of drug, shown to be effective in polyglutamine-induced neurodegeneration and spinal and bulbar muscular atrophy [20, 21] which is potentially useful in treating folding-defective aggregation-prone mutant proteins in the retina. For example, in a cell model of overexpression of P23H opsin, Mendes and Cheetham, [22] showed that 17-AAG prevented opsin protein aggregation and inclusion formation, reducing cell death and caspase activation, while in an *in vivo* murine model of dominant RP, intravitreal administration of 17-AAG (3 μl of 10 mg/ml) also led to a reduction in aggregate formation and cleaved caspase-3 levels resulting in significant protection of the ONL compared to vehicle-only treated mice [19]. Strikingly, in the same study systemic delivery of 17-AAG (30 mg/kg) following modulation of the iBRB by transient downregulation of the tight junction protein, claudin-5, resulted in a similarly high level of ONL protection. Inner retinal barrier modulation was also used by Campbell et al. [23] in the delivery of 17-AAG *i.p.* (31.25 mg/kg) to laser-induced CNV mouse models which resulted in significant reduction in CNV volumes via inhibition of VEGFR2, a client protein of hsp90.

Curcumin, found in turmeric has been used in traditional Indian cuisine and medicine for over 2,000 years. In more recent times curcumin has come to the fore as a pleiotropic therapeutic, targeting inflammation, oxidative stress, and cancer by virtue of the ability to modulate multiple signaling pathways (reviewed in Hatcher et al. [24]) and is currently under clinical trial in 63 different studies (*Clinical-Trials*.gov). Relatively, low bioavailability due to conjugation or reduction in the intestine and bloodstream, respectively, is counterbalanced by clinical tolerance of high doses and has led to research culminating in clinical trials involving curcumin for indications involving neuronal tissue including a Phase III study for Leber's hereditary optic neuropathy and three trials for Alzheimer Disease (reviewed by Gupta et al. [25; Reuter et al. 26]). Curcumin having been shown to reduce protein aggregation and plaque formation in a model of AD [27] was subsequently used by Vasireddy et al. [28] in the P23H rhodopsin rat model (line 1) in which curcumin was delivered by oral gavage @ 100 mg/kg/day from P30 to P70. This resulted in improved retinal morphology, enhanced expression levels of photoreceptor expressed genes, corrected P23H rhodopsin trafficking from ONL to OS, lowered ER stress response protein levels which correlated with protection of ONL thickness to about 60 % that of wild type controls and improved ERGs compared to P23H untreated controls. Bioavailability following oral delivery of curcumin was also measured using liquid chromatography and mass spectrometry. For this, wild type rats received 100 mg/kg curcumin which was subsequently detected in brain and retina at approximately 9 and 28 ng/g respectively 2 h later, indicating that the dose delivered across the BBB and BRB is 10,000 and 3,300 times less respectively than that administered.

Rapamycin, also known as Rapamune/Sirolimus was originally isolated from *Streptomyces hygroscopicus* and developed as an antibiotic but later shown to have many immunomodulatory effects including inhibition of T cell activation and B-lymphocyte proliferation and subsequently gained FDA approval for prevention of transplant rejection and is now also used as an immunosuppressant for uveitis [29]. In addition, rapamycin induces autophagy by complexing to the cytoplasmic protein FKBP12 which then inhibits mammalian target of rapamycin (mTOR), a serine/threonine protein kinase central to cell growth regulation which balances growth with autophagy depending on the status of cellular health and degree of stress. Rapamycin was shown by Kaushal [30] to reduce the presence of P23H but not WT opsin when expressed stably in cell culture and this reduction was associated with autophagy as evidenced by colocalization of markers of autophagy and the inhibition of the process by 3MA, an inhibitor of autophagy. Electron microscopy also revealed the clustering of autophagic vacuoles around P23H opsin protein aggregates. Thus, it was concluded that rapamycin may be a potential therapeutic for retinal diseases associated with protein misfolding such as autosomal dominant P23H rhodopsin RP. However, Cheetham's group, also expressing opsins in a cell culture model, later demonstrated that although rapamycin reduced P23H opsin protein aggregates, inclusion formation, caspase activation and cell death, it did not improve P23H opsin processing or negate the dominant negative effect of the mutant over WT [31]. However, beneficial effects of rapamycin on photoreceptors in vivo have very recently been reported by Kunchithapatham et al. [32]. Using a murine light damage model in which balb-c mice were subjected to 10 or more days of constant bright light (150–175 ft-c) with or without daily rapamycin injections (10 mg/kg daily i.p.), a differential effect of rapamycin on rods and cones was observed. Taken as a unit, photoreceptors appeared protected by rapamycin both structurally and functionally as observed by an increased number of rows of photoreceptors and ERG amplitudes, respectively. However, neither light treatment nor rapamycin had any effect on cone survival indicating that photoreceptor protection by rapamycin is purely due to rod survival. Moreover, closer inspection of scotopic versus photopic ERG responses revealed that while rod function was protected this was not the case for cones since rapamycin actually decreased ERG amplitudes under photopic conditions. Interestingly, however, histological examination showed the presence of autophagic vacuoles in cones and not rods demonstrating that it is in cones that rapamycin selectively induces autophagy. Thus, rapamycin improved rod survival following light damage in an autophagy-independent manner which was then found to be correlated with reduction in levels of the proinflammatory cytokine TGF-β and in VEGF which itself has been previously reported to promote apoptosis in the presence of TGF-β. The autophagy independent, growth factor-lowering effects of rapamycin have great potential for vascular diseases of the retina and already small phase I/II pilot studies found subconjunctival Sirolimus injection to be well tolerated in patients with diabetic macula edema (Dugel et al. [33] (NCT ID: NCT00401115); Krishnadev et al. [34] (NCT ID: NCT00711490)) while a phase II randomized dose-ranging study involving about 120 patients is due for completion in May 2012 (NCT ID: NCT00656643). Trials for both wet and dry AMD using sirolimus (NCT IDs: NCT00766649, NCT01445548, NCT00304954) are also underway or completed but little data indicating efficacy has been published to date.

In regard to the use of small molecules as modulators of oxidative stress and neuroprotection, tauroursodeoxycholic acid (TUDCA) is a hydrophilic bile acid found only in trace amounts in humans but expressed in large quantities in bear. It has, traditionally, been used in Chinese medicine to combat eye disorders and is currently being evaluated in clinical trials for HD, ALS, insulin resistance, transthyretin amyloidosis, cholestasis, and CF (*ClinicalTrials*.gov). Mantopoulos et al. [35] have recently shown that inflammatory and ER stress markers are not suppressed by TUDCA but that reactive oxygen species generated increase in protein carbonyl content was greatly diminished emphasizing that the primary effects of TUDCA are of an antioxidant nature. Boatright et al. [36] were the first to show how potent TUDCA is in preventing retinal degeneration in disease models. *rd10* mice which harbor a point mutation in the β-subunit of rod phosphodiesterase and model recessive RP were administered TUDCA (500 mg/kg) at postnatal days 6, 9, 12, 15, 18 and 30 and exhibited functional and morphological protection including a reduction in the numbers of TUNEL-positive photoreceptors compared to vehicle-treated controls [37]. However, a recent study of similar design demonstrated in *rd10* mice that while photoreceptors were protected at day 30, by day 50 there was a differential effect on rods and cones as assessed by ERG, cones being well protected while rods had degenerated substantially [38]. The authors conclude from these studies that since TUDCA alleviated oxidative stress, as measured by a decrease of superoxide anion production, rod cells unlike cones, may not die due to oxidative stress in this model. Another report where TUDCA was administered weekly (500 mg/kg) to the well researched P23H transgenic rat (line 3) again showed impressive photoreceptor protection compared to controls this time with extensive immunohistological detail showing preservation and high density of pre- and postsynaptic contacts between photoreceptors and bipolar or horizontal cells and preservation of mitochondria in photoreceptors [39].

Nilvadipine is an orally available dihydropyridine L-type calcium channel blocker used as an antihypertensive drug. Suggestions from pilot studies that it reduces the risk of developing cognitive difficulties in hypertensive patients has led to ongoing European-based clinical trials for therapeutic efficacy of the drug in AD patients by the EU-funded Nilvad consortium. Preclinical mechanistic studies show that nilvadipine inhibited $A\beta$ production in vitro while in the double transgenic *PS1/APPsw* mouse model of AD in vivo it reduced $A\beta$ levels in brain and accelerated clearance of $A\beta$ into the plasma across the BBB without altering BBB permeability. Furthermore, chronic oral treatment with nilvadipine improved learning and proficiency at tasks involving spatial memory in these animals [40]. In the visual system Saito et al. [41] noted that despite the failure of the related compound Dilitazem (a non-dihydropyridine L-type calcium channel blocker) to rescue the mutant phenotype in the P23H rhodopsin rat model [42], nilvadipine delivered daily for 4 weeks by i.p. injection (0.25 mg/kg) in line 2 of this model increased ERGs (175 % of untreated P23H control value) and rows of ONL (13 with respect to 8 in control) thus reducing degeneration. The authors show that in this case the mechanism of action of the drug is associated with increased levels of neuroprotective FGF-2 and activity-regulated cytoskeleton-associated protein (Arc), both previously shown to be upregulated in the RCS rat by nilvadipine [43], thus consistent with the preservation of retinal morphology also observed by this group in the same model [44]. Interestingly, Saito et al. [41] further speculate that by blocking L-type (and other) Ca^{2+} channels, nilvadipine may reduce the prolonged duration of phosphorylation of Ser338 and Ser334 in the P23H rat, improving the kinetics of visual cycle quenching of P23H rhodopsin to those of wild type rhodopsin. Finally, nilvadipine has also improved the protection of photoreceptors in other models of RP, namely, heterozygous *RDS* mice in which CNTF and FGF-13 upregulation is a possible mechanism [45] and in *rd1* mice (reviewed by Barabas et al. [46]).

Norgestrel, a synthetic progestin derived from testosterone used as an oral contraceptive and in clinical trial for hypercholesterolemia, hepatitis C, metabolic syndrome, graft vasculopathy, osteoporosis, dyslipidemia, hypertension, coronary disease, renal transplantation and chronic nephropathy, has been shown to improve survival of photoreceptors in two models of retinal degeneration [47]. In the first of these, norgestrel administered to *BALB/c* mice (100 mg/kg *i.p.*) 1 h prior to light ablation (5,000 lux for 2 h) resulted in 75 % reduction in photoreceptor apoptosis in both male and female mice at the 24 h time point. In addition, repeated delivery of norgestrel every 3 days extended protection to 7 days, however, by day 14 apoptosis was widespread. In the second model, *rd10*, (autosomal recessive RP) norgestrel administered every 2 days from P18 preserved photoreceptor number, structure, and function when analyzed at P30. Mechanistic results show that Norgestrel upregulated neuroprotective basic FGF2 in both wild type and *rd10* mice and this correlated in time with downstream activation of prosurvival ERK1/2 signaling.

Looking Toward the Future: Improving the Efficiency with Which Molecular Medicines Can Be Systemically Delivered to the Retina

A very large number of low molecular weight compounds with proven neuro-protective function could, in principle, be used in the prevention or treatment of degenerative retinal diseases, the attraction, as previously emphasized, lying in avoidance of the necessity to directly target specific gene defects. However, in spite of some of the successful indications outlined above, a major limitation has, until recently, been the lack of efficient accessibility of systemically administered drugs to the retina owing to the presence of the inner blood-retina barrier. Recently, however, a relatively straightforward procedure has been described in animal model systems, which greatly facilitates systemic drug delivery to the retina, the efficacy of which has been validated in models of light-induced retinal degeneration and in murine models of autosomal dominant RP and in a laser induced model of the exudative form of age-related macular degeneration.

The molecular architecture of the blood–brain and inner blood-retina barriers are remarkably similar. The retina and brain are suffused with capillaries that are lined by neuronal endothelial cells. These cells are joined together by 'tight-junctions', which are composed of a series of proteins that form an exceedingly tight seal among these cells and it is the tight junctions that form the inner blood-retina barrier (iBRB), preventing passage into the retina from the peripheral circulation of any molecule greater than approximately 200 Da (less than the size of a DNA nucleotide). The iBRB has evolved for a very important reason—to prevent potentially damaging materials, including blood-borne soluble enzymes, byproducts of our immune systems such as C3a and C5a anaphylatoxins, antibodies, and pathogens crossing from the blood supply into the retina. If this barrier is functionally compromised in any way, delicate neuroretinal tissues risk damage by such exposure. Addressing the hypothesis that down-regulation of selected tight junction proteins may render the iBRB more permeable to the passage from the bloodstream of low molecular weight compounds into the retina, while fully maintaining barrier function, Campbell et al. recently began to explore the effects of down regulating tight junction components using RNA interference [48]. The initial target was the TJ component, claudin-5, since this TJ protein is highly enriched in neuronal endothelia. Initially, raw siRNA validated against transcripts encoding claudin-5 was introduced into the tail veins of mice using a high-volume-high-pressure inoculation technique [49]. This resulted in the partial downregulation of claudin-5 protein at the iBRB TJs and as a result, the iBRB became marginally more permeable to systemically deliverable compounds up to a molecular weight of approximately 1kD. This is exactly what the investigators hoped to achieve, a marginal increase in the permeability of the iBRB, without any compromise to the functional integrity of the barrier. Using this technique, barrier permeability could be achieved approximately 24 h post-RNAi administration, the iBRB remaining permeable to low molecular weight compounds for a period of approximately 36 h, during which time siRNA decayed, allowing claudin-5 to reach normal levels. Hence, a 36 h window of permeability was achievable, during which time potentially therapeutic compounds could be systemically administered and gain access to the retina. Campbell et al. were able to demonstrate a rescue of the rod-derived electroretinogram in *IMPDH1-/-* mice by systemic administration of GTP, and a protection of the retinas of mice against light-induced photoreceptor degeneration by systemic delivery of the calpain inhibitor ALLM [49]. In regard to the former, IMPDH1 (inosine monophosphate dehydrogenase 1) is one of two rate limiting enzymes of de novo guanine nucleotide biosynthesis, dominant-acting mutations within the IMPDH1 gene being a known cause of the RP10 form of autosomal dominant RP [50]. In this dominant form of RP, disease pathology is not caused by a lack of guanine nucleotide availability in the retina, rather, by misfolding of the mutated protein [51, 52]. However, mice lacking IMPDH1 also develop a slower retinopathy, where photoreceptor function gradually declines over a period of 12–18 months, although photoreceptors remain largely structurally intact. Campbell et al. were able to demonstrate a recovery in ERG waveforms by systemic administration of GTP post-barrier modulation [49]. They were also able to demonstrate highly efficient protection of the retinas of wild type albino mice against light-induced photoreceptor degeneration. Light-induced degeneration of the retina is a well-known and widely used model of retinal degeneration and is known to be serine protease dependent. Direct intraocular injection of serine protease inhibitors, e.g., ALLM is known to afford protection to murine retinas against light-induced degeneration. Campbell et al. were able to demonstrate highly efficient protection of the retinas of mice in this model, by systemic administration of ALLM following siRNA-mediated barrier modulation. In a subsequent series of experiments, Tam et al. [19] were able to demonstrate protection of the retinas of mice from the effects of expression of a human dominant-

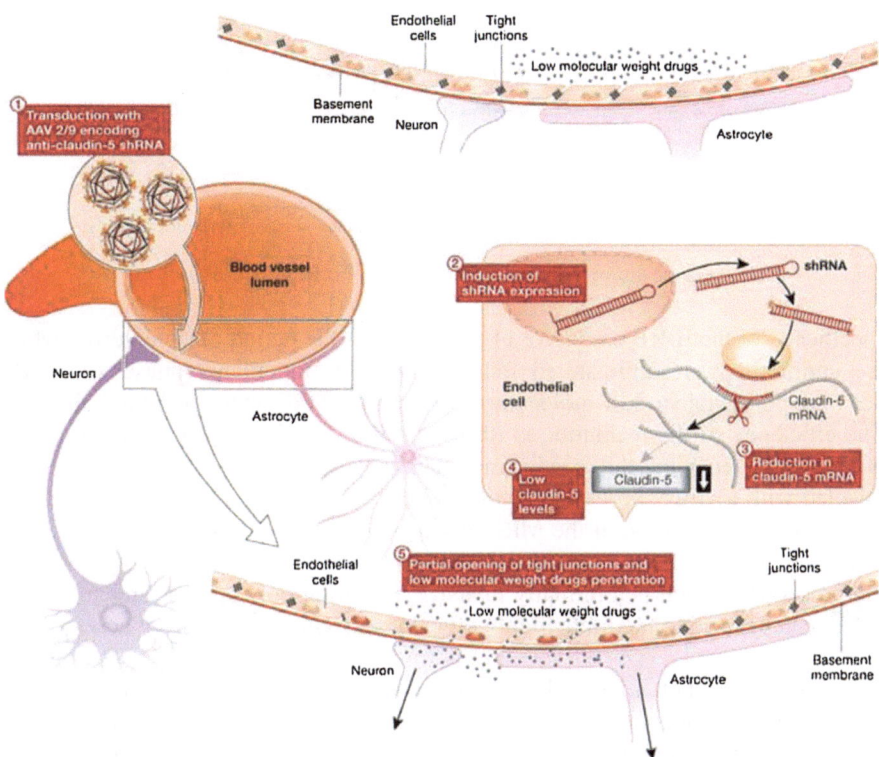

Fig. 3.1 Schematic representation of AAV-mediated neuronal barrier modulation

acting mutation within the IMPDH1 gene (Arg224Pro) by systemic delivery of the Hsp90 chaperone inhibitor, 17-AAG.

For the purposes of human therapy, modulation of the permeability of the iBRB using systemically administered siRNA is not an option. While no adverse effects on the retina or brain were noted in mice using this experimental technique—there were no negative effects on retinal function as assessed by ERG, or on global neuronal transcriptional profiles, administration of large quantities of siRNA into the peripheral circulation, on a regular, e.g., monthly basis, perhaps for many years for the purposes of treatment of chronic disease such as RP, is not a viable proposition. Hence, a technique for localized barrier modulation, limited only to the retina (or indeed to pre-selected regions of the brain) was developed (Fig. 3.1).

According to this technique, a gene encoding shRNA validated in vitro against claudin-5 was introduced into the genome of an AAV2/9 vector such that the gene was under the control of a doxycycline-inducible promoter. The structure of the viral genome is illustrated in Fig. 3.2. AAV2/9 has been shown to transfect neuronal endothelial cells with high efficiency [53]. Such viral particles were inoculated subretinally into the eyes of mice. Administration of doxycycline in drinking water induced shRNA expression, in turn resulting in downregulation of claudin-5

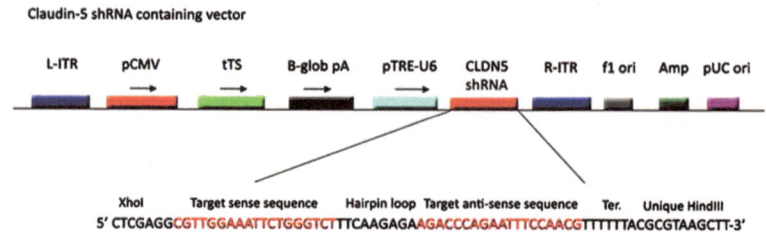

Fig. 3.2   Vector map of doxycycline inducible claudin-5 shRNA construct

specifically at the iBRB exclusive of the brain, for the period of time during which the antibiotic was administered [54]. During this modulated phase, the iBRB becomes permeable to low molecular weight compounds up to approximately 1kD. The efficiency of the technique is illustrated in Fig. 3.3. The right eye of a wild type mouse received the barrier-modulating virus, whereas the left eye was injected with a virus expressing a control shRNA targeting a luciferase gene. The figure shows permeation of the MRI contrasting agent, gadolinium into the retinas of treated animals. This contrasting agent has a molecular weight of 743 Da. Thus, the procedure, which is minimally invasive, requiring only a single inoculation of AAV, can be used to facilitate highly efficient periodic systemic drug access to the retina.

In terms of its potential for human use, Campbell et al. assessed the efficacy of the technique in suppressing choroidal neovascularization in a murine model of the exudative form of age-related macular degeneration (AMD). AMD is the most prevalent cause of registered visual handicap in the developed world. It is a classically multifactorial disease, only rare, early onset forms of the disease seg-regating in a mendelian fashion, established risk factors including cigarette smoking and diet, where macular pigments lutein and zeaxanthin, can only be obtained though vegetable sources, genetic risk factors including components of the complement cascades, including variants in the factor H, factor B (BF), and complement component 2 (C2) genes among others [55–63]. Approximately 80 % of cases of AMD are of the dry, or nonexudative form, where characteristic deposits of proteinaceous debris, termed drusen, accumulate underneath the retina. In early stage disease, a protective role has recently been established for drusen, components of which, including C1q, are known to activate the NLRP3 inflam-masome, resulting in the induction of IL-18 and the suppression of choroidal neovascularization, the characteristic feature of the wet, or exudative form of disease, where new vessels grow from the underlying choroid into the retina, usually resulting in extensive loss of central vision [64]. Many anti-neovascular low molecular weight drugs are known, but have not been used in preemptive treatment of neovascular disease in at risk patients (those showing evidence of drusen deposition) owing to the fact that these drugs cannot with any degree of efficiency, cross the iBRB. Campbell et al. have shown that two such drugs, 17-AAG (also an Hsp90 inhibitor) and sunitinib malate, a vascular endothelial growth factor receptor antagonist, can effectively suppress laser-induced retinal

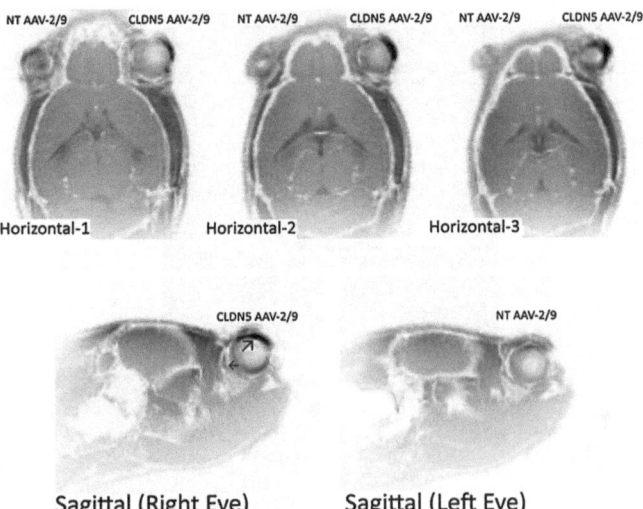

Fig. 3.3 Claudin-5 AAV injected in the right eye and NT AAV injected in the left eye. Gadolinium extravasation is evident in the right eye and not the left eye in a mouse subretinally inoculated with AAV and exposed to doxycycline in drinking water. Top panel: Horizontal orientation, bottom panel, Sagittal orientation

neovascular disease in mice, a widely accepted model of the wet form of AMD ([54], and see Fig. 3.4). In general, systemic delivery of any potentially neuro-protective drug with a molecular weight of up to 1kD, including those outlined above, can be enhanced using this system, with a concomitant minimizing of systemic dosages.

Concluding Remarks

Over the next 10 years, a gradually increasing number of gene replacement therapies for recessive forms of retinopathy will proceed though phase I and II development and enter the clinic. Some of the many forms of experimental gene therapy based on targeting common disease pathologies will also proceed in a similar fashion. Gene-based medicines for dominant forms of retinopathy may also follow, provided that methods targeting dominant-acting mutations prove tolerable to the retina. In parallel, forms of prevention of dominant or recessive retinopathies based upon the use of the growing number of molecular medicines now showing efficacy are also likely to enter the clinic during this timeframe. Given that many of these, again, will target multiple forms of disease, the immense genetic heterogeneity associated with this group of conditions, so long regarded as a daunting barrier, no longer appears to be quite so impenetrable.

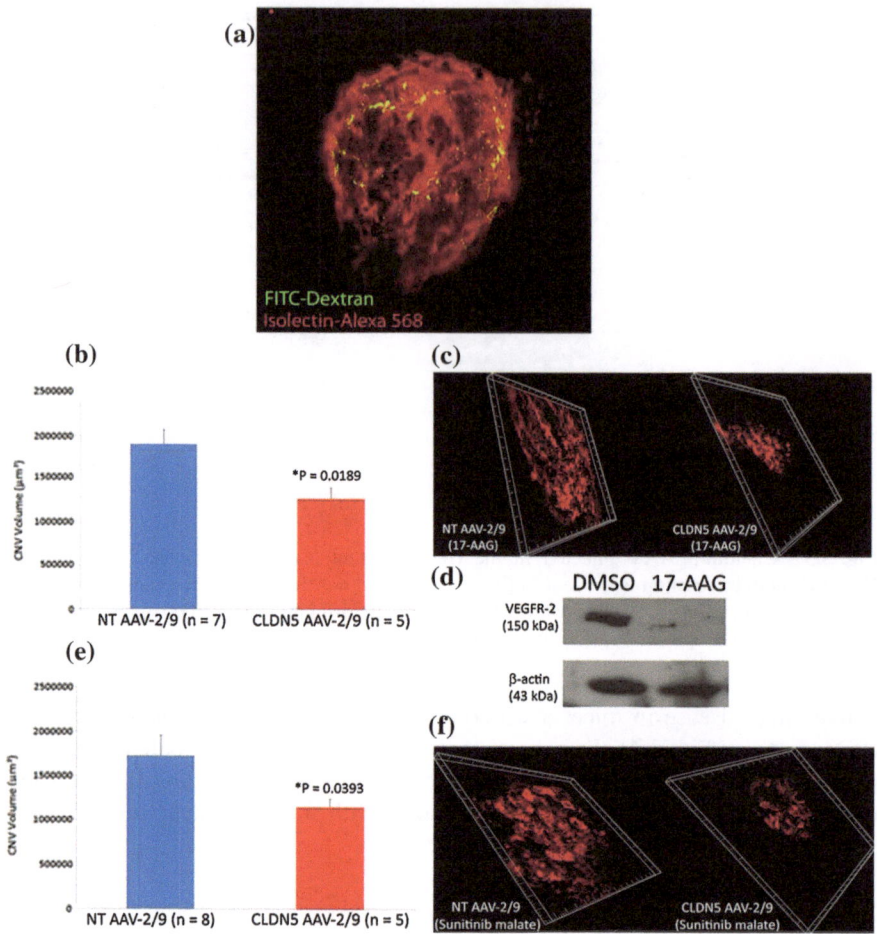

Fig. 3.4   **a** FITC-dextran-200 (FD-200) perfusion of choroidal neovascularization. **b** and **c** Choroidal neovascularization (CNV) was induced with a targeted laser burn (140 mW, 100 mSec, 50 μm spot size) in either the NT AAV-2/9 or CLDN5 AAV-2/9 injected eye and mice were administered 17-AAG intraperitoneally (i.p.). CNV volumes in CLND5 AAV-2/9 injected eyes were significantly reduced when compared to the NT AAV-2/9 injected eyes (*$P = 0.0189$). **d** 17-AAG downregulates VEGFR-2 expression via the inhibition of hsp-90. **e** and **f** Mice were injected with Sunitinib malate (Sutent®) i.p. and CNV volumes were significantly reduced in CLDN5 AAV-2/9 injected mice compared to NT AAV-2/9 controls (*$P = 0.0393$)

References

1. Clemson CM, Tzekov R, Krebs M, Checchi JM, Bigelow C, Kaushal S (2011) Therapeutic potential of valproic acid for retinitis pigmentosa. Br J Ophthalmol 95(1):89–93
2. Berson EL, Rosner B, Sandberg MA, Hayes KC, Nicholson BW, Weigel-DiFranco C, Willett W (1993) A randomized trial of vitamin A and vitamin E supplementation of retinitis pigmentosa. Arch Ophthalmol 111(6):761–772

3. Norton EW (1993) A randomized trial of vitamin A and vitamin E supplementation for retinitis pigmentosa. Arch Ophthalmol 111(11):1460; author reply 1463–1465
4. Marmor MF (1993) A randomized trial of vitamin A and vitamin E supplementation for retinitis pigmentosa. Arch Ophthalmol 111(11):1460–1461
5. Berson EL, Rosner B, Sandberg MA, Weigel-DiFranco C, Moser A, Brockhurst RJ, Hayes KC, Johnson CA, Anderson EJ, Gaudio AR, Willett WC, Schaefer EJ (2004) Clinical trial of docosahexaenoic acid in patients with retinitis pigmentosa receiving vitamin A treatment. Arch Ophthalmol 122(9):1297–1305
6. Berson EL, Rosner B, Sandberg MA, Weigel-DiFranco C, Moser A, Brockhurst RJ, Hayes KC, Johnson CA, Anderson EJ, Gaudio AR, Willett WC, Schaefer EJ (2004) Further evaluation of docosahexaenoic acid in patients with retinitis pigmentosa receiving vitamin A treatment: subgroup analyses. Arch Ophthalmol 122(9):1306–1314
7. Berson EL, Rosner B, Sandberg MA, Weigel-DiFranco C, Brockhurst RJ, Hayes KC, Johnson EJ, Anderson EJ, Johnson CA, Gaudio AR, Willett WC, Schaefer EJ (2010) Clinical trial of lutein in patients with retinitis pigmentosa receiving vitamin A. Arch Ophthalmol 128(4):403–411
8. Massof RW, Fishman GA (2010) How strong is the evidence that nutritional supplements slow the progression of retinitis pigmentosa? Arch Ophthalmol 128(4):493–495
9. Van Hooser PJ, Aleman TS, He Y-G, Cideciyan AV, Kuksa V, Pittler SJ, Stone EM, Jacobson SG, Palczewski K (2000) Rapid restoration of visual pigment and function with oral retinoid in a mouse model of childhood blindness. Proc Natl Acad Sci 97(15):8623–8628
10. Van Hooser PJ, Liang Y, Maeda T, Kuksa V, Jang GF, He YG, Rieke F, Fong HK, Detwiler PB, Palczewski K (2002) Recovery of visual functions in a mouse model of Leber congenital amaurosis. J Biol Chem 277(21):19173–19182
11. Kaushal S, Khorana HG (1994) Structure and function in rhodopsin. Point mutations associated with autosomal dominant retinitis pigmentosa. Biochemistry 33(20):6121–6128
12. Sakami S, Maeda T, Bereta G, Okano K, Golczak M, Sumaroka A, Roman AJ, Cideciyan AV, Jacobson SG, Palczewski K (2011) Probing mechanisms of photoreceptor degeneration in a new mouse model of the common form of autosomal dominant retinitis pigmentosa due to P23H opsin mutations. J Biol Chem 286:10551–10567
13. Bowne SJ, Humphries MM, Sullivan LS, Kenna PF, Tam LC, Kiang A-S, Campbell M, Weinstock GM, Koboldt DC, Ding L, Fulton RS, Sodergren EJ, Allman D, Millington-Ward S, Palfi A, McKee A, Blanton SH, Slifer S, Konidari I, Farrar GJ, Daiger SP, Humphries P (2011) A dominant mutation in RPE65 identified by whole-exome sequencing causes retinitis pigmentosa with choroidal involvement. Eur J Hum Genet 19(10):1074–1081
14. Grover A, Shandilya A, Agrawal V, Pratik P, Bhasme D, Bisaria VS, Sundar D (2011) Blocking the chaperone kinome pathway: mechanistic insights into a novel dual inhibition approach for supra-additive suppression of malignant tumors. Biochem Biophys Res Commun 404:498–503
15. Hertlein E, Wagner AJ, Jones J, Lin TS, Maddocks KJ, Towns WH, Goettl VM, Zhang X, Jarjoura D, Raymond CA, West DA, Croce CM, Byrd JC, Johnson AJ (2010) 17-DMAG targets the nuclear factor-κB family of proteins to induce apoptosis in chronic lymphocytic leukemia: clinical implications of HSP90 inhibition. Blood 116:45–53
16. Rao R, Nalluri S, Fiskus W, Balusu R, Joshi A, Mudunuru U, Buckley KM, Robbins K, Ustun C, Reuther GW, Bhalla KN (2010) Heat shock protein 90 inhibition depletes TrkA levels and signaling in human acute leukemia cells. Mol Cancer Ther 9:2232–2242
17. Tatokoro M, Koga F, Yoshida S, Kawakami S, Fujii Y, Neckers L, Kihara K (2011) Potential role of Hsp90 inhibitors in overcoming cisplatin resistance of bladder cancer-initiating cells. Intl J Cancer. Sep 30 doi:10.1002/ijc.26475
18. Davenport EL, Moore HE, Dunlop AS, Sharp SY, Workman P, Morgan GJ, Davies FE (2007) Heat shock protein inhibition is associated with activation of the unfolded protein response pathway in myeloma plasma cells. Blood 110:2641–2649

19. Tam LCS, Kiang A-S, Campbell M, Keaney J, Farrar GJ, Humphries MM, Kenna PF, Humphries P (2010) Prevention of autosomal dominant retinitis pigmenosa by systematic drug therapy targeting heat shock protein 90 (Hsp90). Hum Mol Genet 19:4421–4436
20. Fujikake N, Nagai Y, Popiel HA, Okamoto Y, Yamaguchi M, Toda T (2008) Heat shock transcription factor 1-activating compounds suppress polyglutamine-induced neurodegeneration through induction of multiple molecular chaperones. J Biol Chem 283:26188–26197
21. Waza M, Adachi H, Katsuno M, Minamiyama M, Sang C, Tanaka F, Inukai A, Doyu M, Sobue G (2005) 17-AAG, an Hsp90 inhibitor, ameliorates polyglutamine-mediated motor neuron degeneration. Nat Med 11:1088–1095
22. Mendes HF, Cheetham ME (2008) Pharmacological manipulation of gain-of-function and dominant-negative mechanisms in rhodopsin retinitis pigmentosa. Hum Mol Genet 17(19):3043–3054
23. Campbell M, Humphries MM, Kiang A-S, Nguyen ATH, Gobbo OL, Tam LCS, Suzuki M, Hanrahan F, Ozaki E, Farrar GJ, Kenna PF, Humphries P (2011) Systemic low-molecular weight drug delivery to pre-selected neuronal regions. EMBO Mol Med 2011(3):235–245
24. Hatcher H, Planalp R, Cho J, Torti FM, Torti SV (2008) Curcumin: from ancient medicine to current clinical trials. Cell Mol Life Sci 65:1631–1652
25. Gupta SC, Patchva S, Koh W, Aggarwal BB (2011) Discovery of curcumin, a component of the golden spice, and its miraculous biological activities. Clin Exp Pharmacol Physiol 39(3):283–299
26. Reuter S, Gupta SC, Park B, Goel A, Aggarwal BB (2011) Epigenetic changes induced by curcumin and other natural compounds. Genes Nutr 6:93–108
27. Yang F, Lim GP, Begum AN, Ubeda OJ, Simmons MR, Ambegaokar SS, Chen PP, Kayed R, Glabe CG, Frautschy SA, Cole GM (2005) Curcumin inhibits formation of amyloid beta oligomers and fibrils, binds plaques, and reduces amyloid in vivo. J Biol Chem 280:5892–5901
28. Vasireddy V, Chavali VRM, Joseph VT, Kadam R, Lin JH, Jamison JA, Kompella UB, Reddy GB, Ayyagari R (2011) Rescue of photoreceptor degeneration by curcumin in transgenic rats with P23H rhodopsin mutation. PLoS ONE 6(6):e21193. doi:10.1371/journal.pone.0021193
29. Shanmuganathan VA, Casely EM, Raj D, Powell RJ, Joseph A, Amoaku WM, Dua HS (2005) The efficacy of sirolimus in the treatment of patients with refractory uveitis. Br J Ophthalmol 89:666–669
30. Kaushal S (2006) Effect of rapamycin on the fate of P23H opsin associated with retinitis pigmentosa. Trans Am Ophthalmol Soc 104:517–529
31. Mendes HF, Zaccarini R, Cheetham ME (2010) Pharmacological manipulation of rhodopsin retinitis pigmentosa. Adv Exp Med Biol 664:317–323
32. Kunchithapautham K, Coughlin B, Lemasters JJ, Rohrer B (2011) Differential effects of rapamycin on rods and cones during light-induced stress in albino mice. Invest Ophthalmol Vis Sci 52:2967–2975
33. Dugel PU, Blumenkranz MS, Haller JA, Williams GA, Solley WA, Kleinman DM, Naor J (2012) A randomized, dose-escalation study of subconjunctival and intravitreal injections of sirolimus in patients with diabetic macular edema. Ophthalmology 119(1):124–131
34. Krishnadev N, Forooghian F, Cukras C, Wong W, Saligan L, Chew EY, Nussenblatt R, Ferris F 3rd, Meyerle C (2011) Subconjunctival sirolimus in the treatment of diabetic macular edema. Graefes Arch Clin Exp Ophthalmol 249(11):1627–1633
35. Mantopoulos D, Murakami Y, Comander J, Thanos A, Roh M, Miller JW, Vavvas DG (2011) Tauroursodeoxycholic acid (TUDCA) protects photoreceptors from cell death after experimental retinal detachment. PLoS ONE 6(9):e24245. doi:10.1371/journal.pone.0024245
36. Boatright JH, Moring AG, McElroy C, Phillips MJ, Do VT, Chang B, Hawes NL, Boyd AP, Sidney SS, Stewart RE, Minear SC, Chaudhury R, Ciavatta VT, Rodrigues CMP, Steer CJ, Nickerson JM, Pardue MT (2006) Tool from ancient pharmacopoeia prevents vision loss. Mol Vis 12:1706–1714

37. Phillips MJ, Walker TA, Choi H-Y, Faulkner AE, Kim MK, Sidney S, Boyd AP, Nickerson JM, Boatright JH, Pardue MT (2008) Tauroursodeoxycholic acid preserves photoreceptor structure and function in the rd10 mouse through post-natal day 30. Invest Ophthalmol Vis Sci 49(5):2148–2155
38. Oveson BC, Iwase T, Hackett SF, Lee SY, Usui S, Sedlak TW, Snyder SH, Campochiaro PA, Sung JU (2011) Constituents of bile, bilirubin and TUDCA, protect against oxidative stress-induced retinal degeneration. J Neurochem 116:144–153
39. Fernandez-Sanchez L, Lax P, Pinilla I, Martin-Nieto J (2011) Tauroursodeoxycholic acid prevents retinal degeneration in transgenic P23H rats. Invest Ophthalmol Vis Sci 52:4998–5008
40. Paris D, Bachmeier C, Patel N, Quadros A, Volmar C-H, Laporte V, Ganey G, Beaulieu-Abdelahad D, Ait-Ghezala G, Crawford F, Mullan MJ (2011) Selective antihypertensive dihydropyridines lower $A\beta$ accumulation by targeting both the production and the clearance of $A\beta$ across the blood-brain barrier. Mol Med 17:149–162
41. Saito Y, Ohguro H, Ohguro I, Sato N, Yamazaki H, Metoki T, Ito T, Nakazawa M (2008) Misregulation of rhodopsin phosphorylation and dephosphorylation found in P23H rat retinal degeneration. Clin Ophthalmol 2:821–828
42. Bush RA, Kononen L, Machida S, Sieving PA (2000) The effect of calcium channel blocker diltiazem on photoreceptor degeneration in the rhodopsin Pro23His rat. Invest Ophthalmol Vis Sci 41:2697–2701
43. Sato M, Ohguro H, Ohguro I, Mamiya K, Takano Y, Yamazaki H, Metoki T, Miyagawa Y, Ishikawa F, Nakazawa M (2003) Study of pharmacological effects of nilvadipine on RCS rat retinal degeneration by microarray analysis. Biochem Biophys Res Commun 306:826–831
44. Yamazaki H, Ohguro H, Maeda T, Maruyama I, Takano Y, Metoki T, Nakazawa M, Sawada H, Dezawa M (2002) Preservation of retinal morphology and functions in royal college surgeons rat by nilvadipine, a Ca(2 +) antagonist. Invest Ophthalmol Vis Sci 43:919–919
45. Takeuchi K, Nakazawa M, Mizukoshi S (2008) Systemic administration of nilvadipine delays photoreceptor generation of heterozygous retinal degeneration slow (rds mouse). Exp Eye Res 2008(86):60–69
46. Barabas P, Peck CC, Krizaj D (2010) Do calcium channel blockers rescue dying photoreceptors in the $Pde6b^{rd1}$ mouse? Adv Exp Med Biol 2010(664):491–499
47. Doonan F, O'Driscoll C, Kenna P, Cotter TG (2011) Enhancing survival of photoreceptor cells in vivo using the synthetic progestin. Norgestrel J Neurochem 2011(118):915–927
48. Campbell M, Kiang AS, Kenna PF, Kerskens C, Blau C, O'Dwyer L, Tivnan A, Kelly JA, Brankin B, Farrar GJ, Humphries P (2008) RNAi-mediated reversible opening of the blood-brain barrier. J Gene Med 10(8):930–947
49. Campbell M, Nguyen AT, Kiang AS, Tam LC, Gobbo OL, Kerskens C, Ni Dhubhghaill S, Humphries MM, Farrar GJ, Kenna PF, Humphries P (2009) An experimental platform for systemic drug delivery to the retina. Proc Natl Acad Sci U S A 106(42):17817–17822
50. Kennan A, Aherne A, Palfi A, Humphries MM, Stitt A, Simpson D, Demtroder K, Orntoft T. Ayuso C, Kenna PF, Farrar GJ, Humphries P (2002) Identification of an IMPDH1 mutation in autosomal dominant retinitis pigmentosa (RP10) revealed following comparative microarray analysis of transcripts derived from retinas of wild-type and Rho-\- mice. Human Mol Genet 11(5):547–558
51. Aherne A, Kennan A, Kenna PF, McNally N, Lloyd DG, Alberts IL, Kaing A-S, Humphries MM, Ayuso C, Engel PC, Gu JJ, Mitchell BS, Farrar GJ, Humphries P (2004) On the molecular pathology of neurodegeneration in IMPDH1-based retinitis pigmentosa. Hum Mol Genet 13(6):641–650
52. Wang XT, Mion B, Aherne A, Engel PC (2011) Molecular recruitment as a basis for negative dominant inheritance? Propagation of misfolding in oligomers of IMPDH1, the mutated enzyme in the RP10 form of retinitis pigmentosa. Biochim Biophys Acta 1812(11):1472–1476
53. Foust KD, Nurre E, Montgomery CL, Hernandez A, Chan CM, Kaspar BK (2009) Intravascular AAV9 preferentially targets neonatal neurons and adult astrocytes. Nat Biotechnol 27(1):59–65

54. Campbell M, Humphries MM, Nguyen ATH, Gobbo OL, Tam LCS, Suzuki M, Hanrahan F, Ozaki E, Farrar G-J, Kiang A-S, Kenna PF, Humphries P (2011) Systemic low molecular weight drug delivery to pre-selected neuronal regions. EMBO Mol Med 3:235–245

55. Seddon JM, George S, Rosner B (2006) Cigarette smoking, fish consumption, omega-3 fatty acid intake, and associations with age-related macular degeneration: the US Twin Study of Age-Related Macular Degeneration. Arch Ophthalmol 124(7):995–1001

56. Chen Y, Bedell M, Zhang K (2010) Age-related macular degeneration: genetic and environmental factors of disease. Mol Interv 10(5):271–281. Review

57. Klein RJ, Zeiss C, Chew EY, Tsai JY, Sackler RS, Haynes C, Henning AK, SanGiovanni JP, Mane SM, Mayne ST, Bracken MB, Ferris FL, Ott J, Barnstable C, Hoh J (2005) Complement factor H polymorphism in age-related macular degeneration. Science 15;308(5720):385–389

58. Edwards AO, Ritter R 3rd, Abel KJ, Manning A, Panhuysen C, Farrer LA (2005) Complement factor H polymorphism and age-related macular degeneration. Science 15;308(5720):421–424

59. Haines JL, Hauser MA, Schmidt S, Scott WK, Olson LM, Gallins P, Spencer KL, Kwan SY, Noureddine M, Gilbert JR, Schnetz-Boutaud N, Agarwal A, Postel EA, Pericak-Vance MA (2005) Complementy factor H variant increases the risk of age-related macular degeneration. Science 308(5720):419–421

60. Gold B, Merriam JE, Zernant J, Hancox LS, Taiber AJ, Gehrs K, Cramer K, Neel J, Bergeron J, Barile GR (2006) AMD Genetics Clinical Study Group, Hageman GS, Dean M, Allikmets R. Variation in factor B (BF) and complement component 2 (C2) genes is associated with age-related macular degeneration. Nat Genet 38(4):458–462

61. Edwards AO, Malek G (2007) Molecular genetics of AMD and current animal models. Angiogenesis 10(2):119–132 Review

62. Canter JA, Olson LM, Spencer K, Schnetz-Boutaud N, Anderson B, Hauser MA, Schmidt S, Postel EA, Agarwal A, Pericak-Vance MA, Sternberg P Jr, Haines JL (2008) Mitochondrial DNA polymorphism A4917G is independently associated with age-related macular degeneration. PLoS ONE 3(5):e2091

63. Leveziel N, Tilleul J, Puche N, Zerbib J, Laloum F, Querques G, Souied EH (2011) Genetic factors associated with age-related macular degeneration. Ophthalmologica 226(3):87–102

64. Doyle SL, Campbell M, Ozaki E, Salomon RG, Mori A, Kenna PF, Farrar GJ, Kiang AS, Humphries MM, Lavelle EC, O'Neill LA, Hollyfield JG, Humphries P (2012) NLRP3 has a protective role in age-related macular degeneration through the induction of IL-18 by drusen components. Nat Med 8 Apr doi:10.1038/nm.2717. [Epub ahead of print]